Maverick
GARDENERS

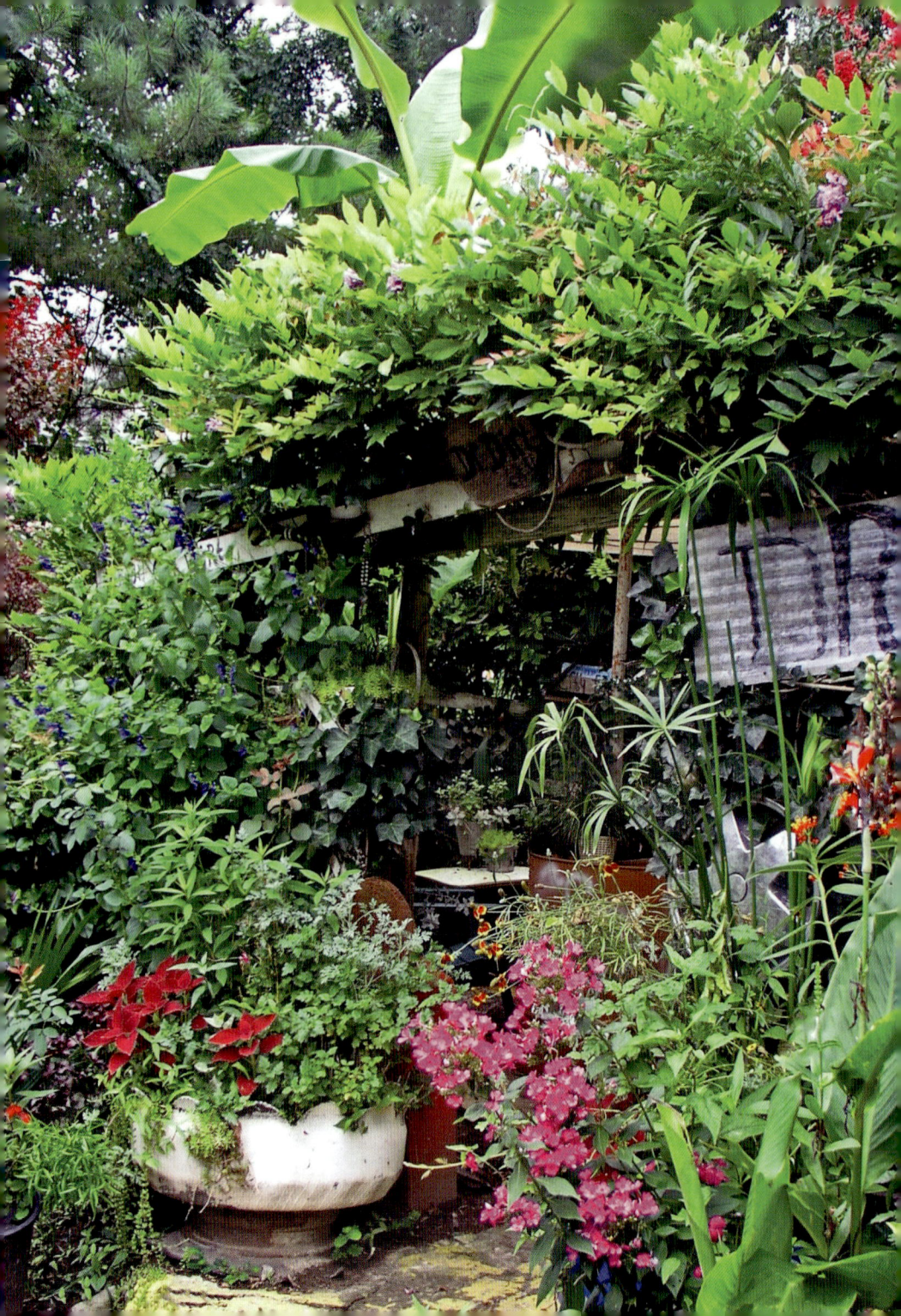

Maverick GARDENERS

Dr. Dirt and Other Determined Independent Gardeners

Felder Rushing

University Press of Mississippi / Jackson

The University Press of Mississippi is the scholarly publishing agency of the Mississippi Institutions of Higher Learning: Alcorn State University, Delta State University, Jackson State University, Mississippi State University, Mississippi University for Women, Mississippi Valley State University, University of Mississippi, and University of Southern Mississippi.

www.upress.state.ms.us

The University Press of Mississippi is a member of the Association of University Presses.

Copyright © 2021 by Felder Rushing
All rights reserved
Manufactured in Korea

All photographs are courtesy of the author unless otherwise noted.

First printing 2021

∞

Library of Congress Cataloging-in-Publication Data available

LCCN 2020042497
ISBN 9781496832214 (hardback)
ISBN 9781496832719 (trade paperback)
ISBN 9781496832726 (epub single)
ISBN 9781496832733 (epub institutional)
ISBN 9781496832740 (pdf single)
ISBN 9781496832757 (pdf institutional)

British Library Cataloging-in-Publication Data available

This book is for self-reliant people worldwide who lovingly tend personally meaningful gardens for the joy of it—especially those who feel isolated because they are okay with coloring outside the lines.

Contents

Acknowledgments ix

Introduction 3

1 Different Peas, Same Pod 5

2 Gardener Coming Out 21

3 A Man Called Dirt 39

4 More Keepers of the Flame: DIGr Neighbors 83

5 Plants: Growing It Forward 125

6 Yard Art: The Good, the Bad, and the Unbelievable 141

7 Putting It Together 159

Coda: DIGrs Infinity 179

Empowering Terms for DIGrs 181

Acknowledgments

I deeply appreciate each of the gardening friends who have encouraged me, by example and usually with good humor.

There were the four generations of family gardening women who raised me in their disparate gardens (more on them later); I'm still sharing some of their plants and wisdom, including with my own daughter, Zoe.

I am very lucky to have friends scattered across the country who, like me, are conflicted horticulturists and journalists grappling with ways to reach our audiences with a balance between exacting science and the less precise, corner-cutting "make-do" approach followed blissfully by garden variety gardeners.

Most of us tease, torment, and yet encourage one another in our struggles. I started to list a few, but got bogged down in sweet memories, and the list got too long. You know who you are, and most of you know one another.

But I must give a long-overdue memorial nod to Flora McQuirter, who gardened by herself with her dogs along the Sunflower River that winds slowly through the heart of the Mississippi Delta. When Miss Flora occasionally came to town to get supplies, she would stop by the garden center where I worked as a teenager and give me cuttings of unusual plants from her garden to share with others. Flora, who casually cussed a lot, was the first "crazy cat lady" gardener in my life.

Special shout-outs go to the Cottage Garden Society of England and PlantSwap UK for their collective centuries of insights, and to Janice Watkins, the retired Flora, Mississippi, librarian who in 1990 mixed dozens of diverse independent gardeners and countless precious plants into what is now the longest-running plant swap in the known universe.

This book is dedicated to all these and more, along with the folks at University Press of Mississippi who fervently believe in the merits of bringing stuff like what's in this book to light.

And finally, I am grateful to Susan, who nudged and tweaked this book from being a mish-mash rave of dry observations into a sweet waltz.

Knowing that Nature never did betray
The heart that loved her; 'tis her privilege,
Through all the years of this our life, to lead
From joy to joy . . .
—William Wordsworth

Maverick
GARDENERS

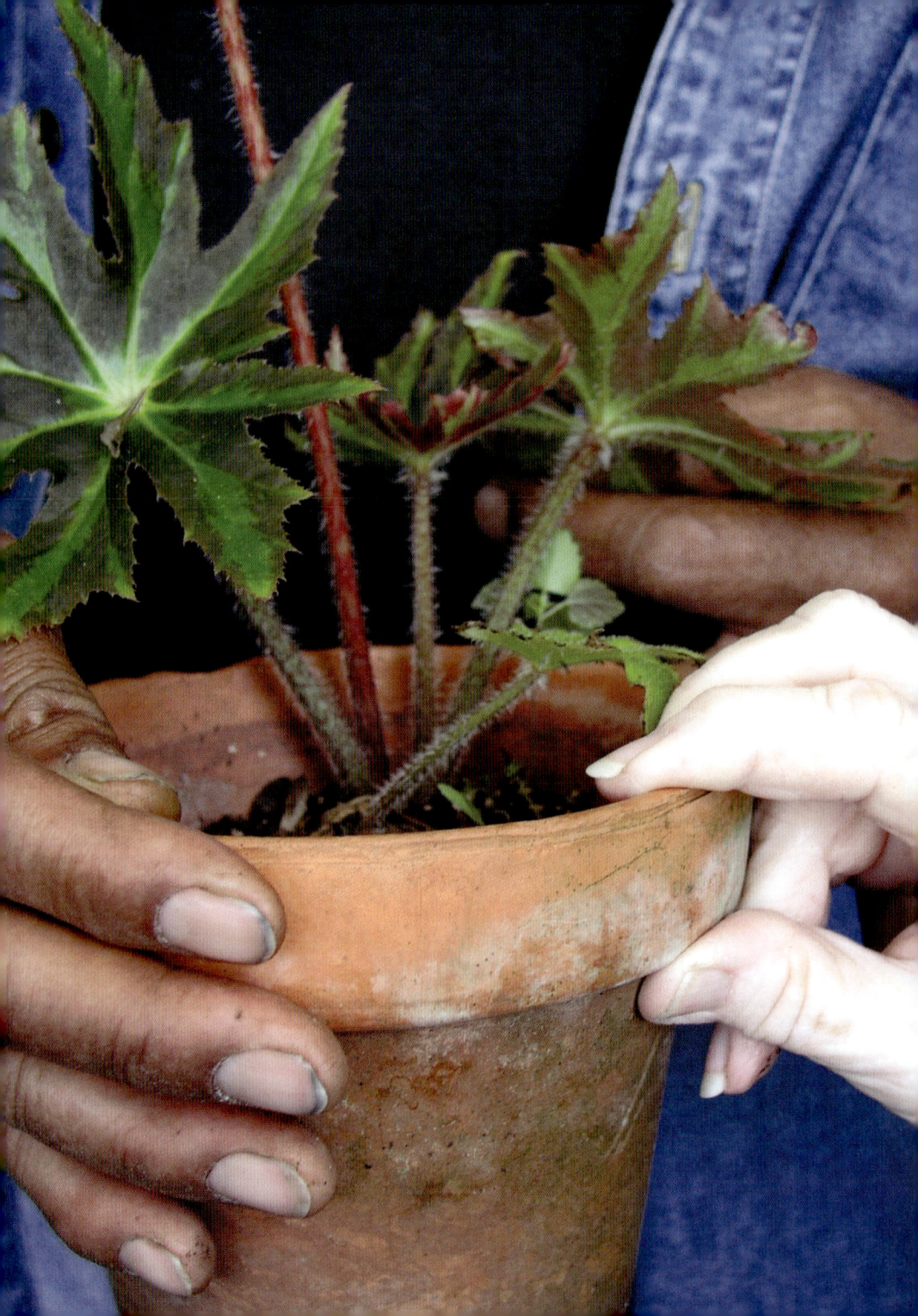

Introduction

> There is no such thing as a weird human being. It's just that some people require more understanding than others.
> —**Tom Robbins**

Somebody told me I ought to write a book about the quirky "odd sock" garden extemporizers I lovingly portray in my lectures. A book featuring those rarely celebrated real people who weave and wobble a precarious line between sanity and letting go is long overdue.

To be clear, most of these gardeners are not rebellious nonconformists; they are merely other-motivated.

So here goes. Be forewarned that this book honors people like the woman in my hometown who paints the numbers of her favorite NASCAR drivers on her elephant ears, and a Tokyo gardener with over a hundred bonsai plants. I met a woman in rural Devon, England, with two thousand gnomes playing amongst the vegetation, and an African desert gardener who created an imaginary front walk by lining a carefully swept dirt path with upended wine bottles.

These gardeners matter, whether or not their efforts are acknowledged or appreciated.

Plants don't care who you are.

You may already be an old hand at finding, nurturing, and sharing plants. If not, you will find more than enough "how-to" covered in most other gardening books and online, with enough detail to make your eyes bleed. Heck, I've written a couple dozen or more myself.

This book is more about those gardeners who obsess over plants to the point where they stand out from others, usually spectacularly. The ones who know that plants have their own special kind of souls and that weeds are just vulnerable plants in need of an appropriate spot to shine.

This is by no means a book about just Southern gardeners, though I am deliberately featuring a handful from within easy walking distance of my own cottage in the small village of Fondren, Mississippi. With their interwoven stories and experiences, these few exemplify all the others scattered across the world, from California and Boston to England and Japan, who, while not always tending the same plants, certainly trod and till similar patches of soil.

They lovingly mind their home grounds, sustainably, without trying to provide a global solution, understanding that, as Steve Bender put it, "Maybe you can't change the whole world. But you can make little changes in your own backyard, and that's a start."

Perhaps you are one of these determined, independent gardeners, perhaps not. But I bet you know someone who is, and, truth be told, you'd be better off for it. It's an umami flavor of life thing.

This book, then, is about those Keepers of the Garden Flame who, in the face of misunderstanding and sometimes outright mockery, nurture our tribal knowledge of gardening for the love of it and carry overlooked plants and invaluable lore forward.

1

Different Peas, Same Pod

> The world to me is a secret which I desire to divine. Curiosity, earnest research to learn the hidden laws of nature, and a gladness akin to rapture are among the earliest sensations I can remember.... There is a love of the marvelous—a belief in the marvelous—which hurries me out of the common pathway of other men.
> —**Mary Shelley,** Dr. Victor Frankenstein's self-description

It takes all sorts. As goes the saying in my second home in Lancashire, northern England, about the curious things some people do: "There's nowt so queer as folk."

Think about how we unconsciously take a second look at those earnest celebrants who seasonally gaudy-up their homes and gardens with holiday lights and figures. We are drawn like moths to flames, as if we see something of ourselves we suppress in them.

Take it a step farther: There are overly zealous gardeners scattered lightly across all cultures who are of much the same character when it comes to overplanting and over-accessorizing.

What's up with these people?

Unconventional barely touches their unique, experiential approach as they, innocently or deliberately, push back against rigid thinking and enforced ideals.

Unfortunately or not, many of them don't seem to care that their visions don't gel with those of neighbors, and that they are often excluded from formal gardening circles—or worse, are tacitly or even openly ridiculed.

Yet in some ways they are the most honest gardeners. They have a childlike *tabula rasa* attitude, as if they are learning everything anew through experience and perception. They are open and ever changing, and mind their own business while questioning the local dogma of what is acceptable.

A friend who wrote for the *New York Times* once suggested to me that these dedicated independents are simply garden lovers, amateurs being from Latin *amāre*, to love.

Earnest, not Rebellious

Not to say they are stubborn nonconformists; far from being deliberately rebellious, they are often merely going their own sweet ways. Mostly they just garden so passionately they veer away from the norm. It's in them and has to come out. And though there's often a bit of good humor, they're not trying to be funny.

Many of these gardeners are nonjoiners at heart—loners who sometimes have trouble finding suitable venues for sharing with one another, yet are uncomfortable in organized settings. They are usually most relaxed just tooling around the yard by themselves or in the occasional company of a few likeminded friends or eager visitors.

Some may belong to a garden club or plant group or participate in the local Master Gardener program, but most don't feel entirely comfortable hewing to "proper" rules of horticulture, let alone kowtowing to the inevitable social aspects of most organizations.

Deep down they may be unsocial out of a lack of need; some are downright cantankerous, bored by the same old same old. Those that actually get involved with groups usually end up on the plant swap committee. The main thing they all have in common is a love of flowers, sharing, and accessorizing.

DIGrs Defined

For years, I referred to these "brothers and sisters of the spade" as *dirt gardeners.* But when I asked listeners to my radio program for a broader term, Bill Thames, from Laurel, Mississippi, coined a simple but brilliant acronym: DIGrs—for Determined Independent Gardeners.

These earnest outlier gardeners are not to be confused with the seventeenth-century English religious movement called True Levellers, widely known as The Diggers because of their radical attempts to farm illegally on common land.

And though they garden alone, these seeming outliers are not alone in what they do; whether they know it or not, they are actually a

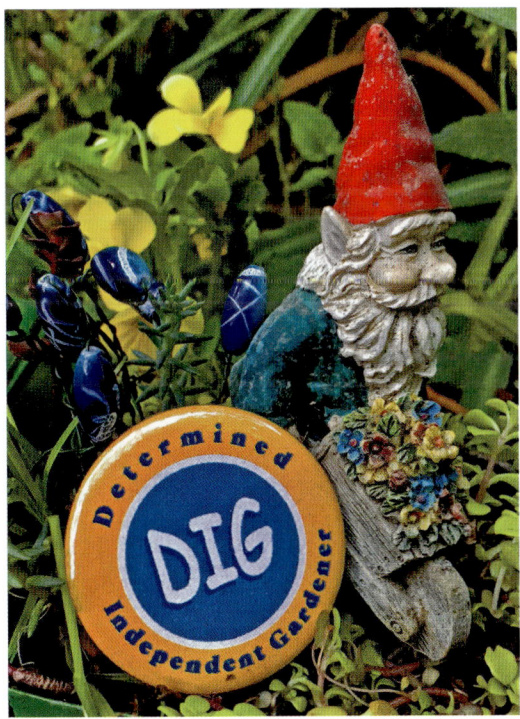

loosely affiliated tribe bound by plants and attitude, scattered around town and countryside.

Oh, you'll find them at informal plant sales, or sitting in the back rows at garden lectures, and maybe prowling around botanic gardens looking for ideas. And if you were to pop by their gardens, they usually are very welcoming and glad to chat, and insist on your taking a homegrown plant or two.

They are the mostly likely gardeners to understand that the rules of horticulture are tempered with certain inalienable rights such as being free to display as many wind chimes and gnomes as they want, plant any color flower next to any color flower, and try (sometimes successfully) doing things the experts say won't work.

Different Drummers

See, it's a personality type that just happens to express itself through gardening. The same kinds of people can be found exposing their inner selves via other outlets, including the way they dress or collect memorabilia. We've invented sayings for them: they march to a different drum, are square pegs in round holes, or are a different kettle of fish; however we put it, they don't always do what society prescribes for them.

I've met DIGrs in their private oases scattered somewhat randomly across five continents. Perhaps there is one in your own family or neighborhood. If not, just inquire at any local garden center or Master Gardener group. Or simply drive around and you'll usually find one in nearly every neighborhood. Heck, ask your postal delivery person who probably sees several in daily rounds.

More than anywhere else, I have met DIGrs at plant swaps which are typically Central Station for folks who love sharing and getting plants without becoming too involved.

Slow Gardening

In a blowback against outsourced mow-and-blow lawn care and other instant gratification or fast-food-style gardening, I honed in my *Slow Gardening* book the concept of a more personalized gardening style which, more than a mere checklist for easy gardening, is close kin to relaxed home cooking. It's an attitude of thinking long haul and enjoying what you do.

Slow gardening is a big tent under which many different personalities and styles can coexist, including gardeners who grow vegetables, herbs, flowers, and fruits, tend their own lawn, or have an intense garden hobby such as topiary, bonsai, or plant hybridizing. It even embraces related interests such as weather watching, garden photography, interest in all types of wildlife, and visiting other gardens.

Slow gardening is by no means lazy or passive; it often involves doing more stuff, carefully selected to be satisfying or productive without senseless, repetitive chores. Its participants savor everything they do, using all senses through all seasons.

By focusing on seasonal rhythms and local conditions, it helps the gardener get more from the garden while better appreciating how leisure time and energy are spent.

Slow gardeners are alert to how things look, sound, smell, feel, and taste—including the not-so-pleasant. You know that fingertip tingle after scrubbing dirt from under your nails? And using a stick to get mud off your boots? And the pungent smell of partly composted stuff? *Those are parts of gardening, too.*

The slow approach can help us fully grasp the little things in our gardens all year. A few ways to practice slow gardening:

- Spread out your chores, doing a little as you go.
- Design in something for all the senses, including emotions.

- Develop a repetitive or year-round hobby such as bonsai, beekeeping, or bird watching.
- When practical, use quiet hand tools over noisy machines.
- Get personal with weather with a rain gauge and thermometer.
- Use solar energy to make tea and dry clothes (scratchy is good).
- Install a cheery firepit and a water feature with a splashy waterfall; lengthen your garden swing's chain for a slower arc.
- Get to know local growers at a farmers' market.
- Take advantage of garden lectures and plant swaps.
- Keep a garden journal, and share seasonal photos online.

More than a mere physical place, a garden is a concept of possibility or potential. The gardener who chooses between different possibilities from moment to moment, and slows down to appreciate the effort as well as the results, ends up with a more personal and satisfying garden.

Love of Life

In small gardens, close-at-hand plant and animal cycles are compressed and fast-paced, making them more accessible and intense which hones our natural, sustaining love for life. This compels most of us to seek connections with other forms of life, be they pets or potted plants, or by taking a walk through the neighborhood.

The concept of *biophilia*, of finding benefits and happiness through sharing a "love of life" with others, was around in Aristotle's day. It helped explain why ordinary people care for and sometimes risk their lives to save wild animals, and keep plants and flowers in and around their homes.

Can you tell which is the Slow Gardener?

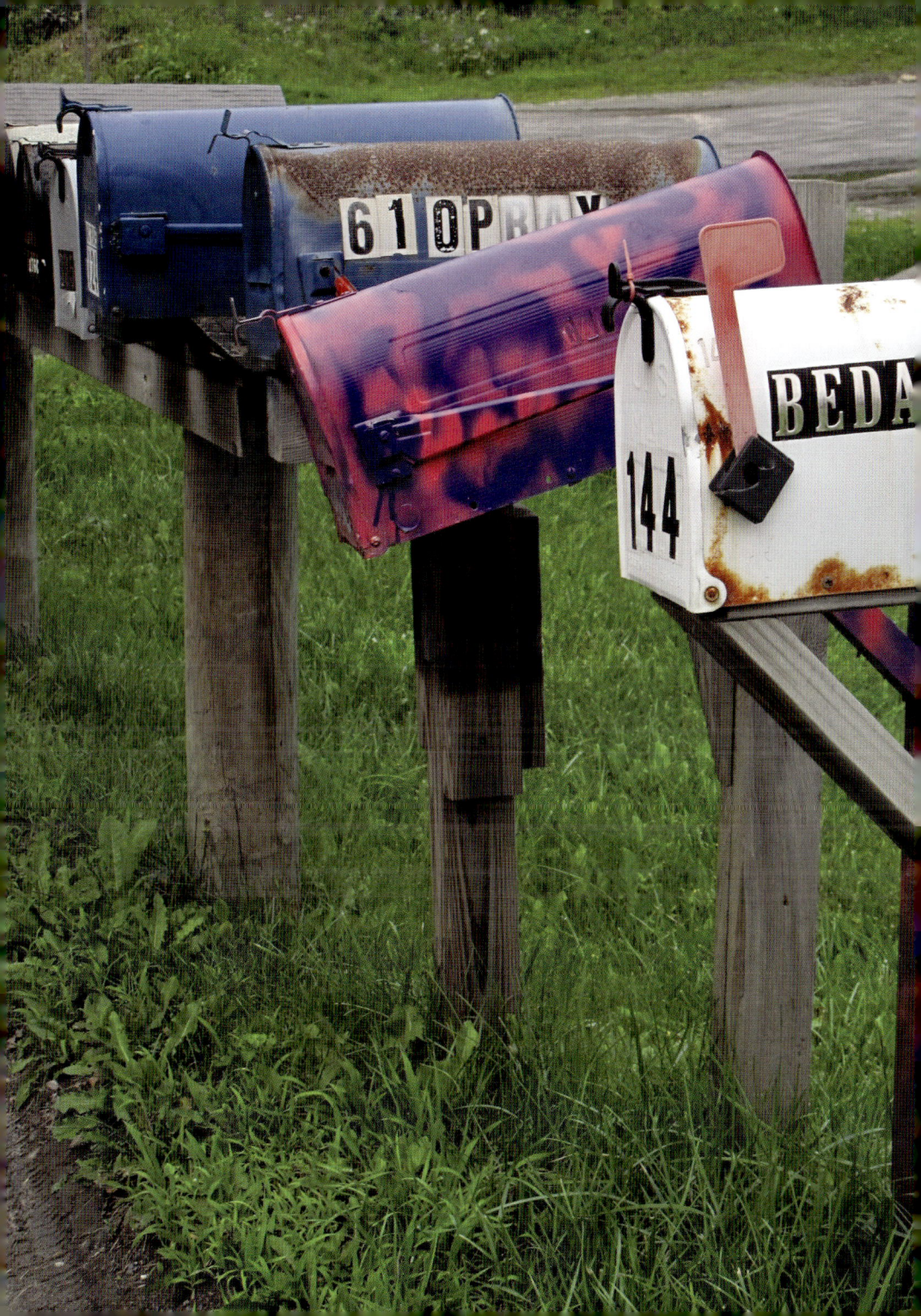

These connections are crucial in modern days because of rapidly increasing technology and time spent cooped up indoors or in our cars disconnected from nature. This detachment helps explain the resurgent urge among people to reconnect with nature, as evidenced by their activism in climate change issues.

This is where DIGrs, with their natural love of life, come in.

Many hard-core DIGrs see weeds as metaphors for themselves; when someone calls a beautiful wildflower in their gardens a weed, they interpret it as saying their garden itself is a weed in the community.

Rather than mere collectors, they are mostly gatherers and sharers, tolerant of and embracing imperfections; they delight in the everyday, making every day a delight.

This actually puts most DIGrs at the top of their psychological game. Like anyone striving to rise above their situation, gardeners fall somewhere along Abraham Maslow's classic hierarchy of needs, which starts with the very basics of survival and builds on security, then social-related needs, while striving to reach the level where they are completely free from worry.

They go from simple hand-to-mouth planting whatever they can and hoping to keep it alive, to improving their soil and techniques and saving seed for the next year. Once these skills are assured, they start growing a wider selection and accessorizing for beauty, and eventually form alliances with other gardeners for support with plants and information.

By the time gardeners reach this level, they already know how to nurture a good variety of plants and propagate enough to share with a wider group. They become teachers of others as well.

Those who transcend all have reached the most relaxed level called *self-actualization*, the nonplussed point where they are okay just doing their own thing and gladly share with anyone who approaches them in a similar hassle-free spirit.

Celebrating the Peculiar

Miss Universe, move over—there's another way to judge beauty that makes as much sense. Maybe more.

Koten engei is the Japanese celebration of slight imperfections, in which critical importance is bestowed upon gnarly and bizarre mutations and deviations. The more naturally twisted, variegated, spotted, and frilled a plant is, the higher its score.

Think "ugliest dog" contests.

It's a celebration of roughness called *wabi-sabi* in which asymmetry, impermanence, and the incomplete, even sadness, are appreciated as reminders of the passage of time and transience of life.

It's deliberately leaving a dead branch on a bonsai plant, or cultivating ferns and mosses in a stumpery garden, or Spanish moss in the kitchen window. Hanging a weathered window frame on a wall or fence. Using broken crockery as a flower border, or placing Granny's chipped old urn in the iris bed.

DIGrs understand this, perhaps subliminally; next time you feel inclined to criticize their love of tree-hugging epiphyte plants, think of it as amateur koten engei.

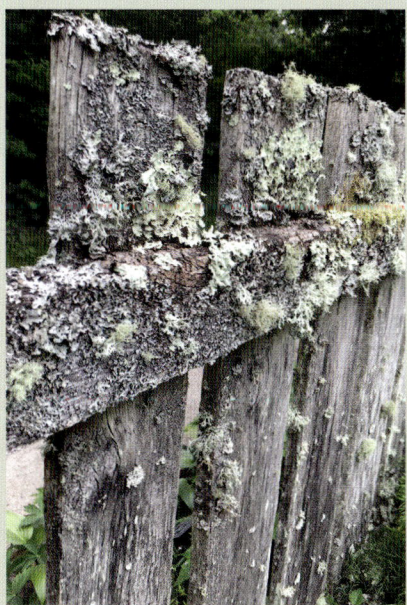

Plant beauty—eye of the beholder thing.

This is the rarified level on which most DIGrs can be found. They share two emotions: proud (sometimes doggedly so) to be true to themselves and hopeful that others will be understanding, or at least cut them some slack. Their mantra is "It's not that I don't care what others think—*it just doesn't matter*!"

No, thank you, they don't need a twelve-step program. Not hurting anyone, any more than do those folks who dress in costumes and smear on greasepaint to go cheer for their favorite sports team.

And oddly, feeling a little like outsiders makes DIGrs all the more welcoming to those who inquire within; *they're pleased to have you in their gardens.*

Origins of the "Green Thumb"

For a long time, I had no idea what made me click with unusual aspects of gardening, and wanted to explain them. Then I read *Frames of Mind: The Theory of Multiple Intelligences*, in which Harvard psychology professor Howard Gardner postulated that most of us are born with at least nine major types of intelligences, developed more or less in each person. Examples are musical, linguistic, mathematical, and athletic.

The one he called "naturalist" has to do with how people see subtle patterns in their natural surroundings, are inspired and rejuvenated by them, and use these abilities productively.

Those with higher naturalist intelligence typically notice faint distinctions and nuances in their surroundings. These abilities give them an edge in hunting, fishing, forecasting weather, watching birds, managing an aquarium, collecting rocks, managing wildlife or wildflowers, adding the right spices to home-cooked meals . . . and gardening. It's what helps some people easily spot arrowheads, four-leaf clovers, and heart-shaped rocks. They can sense if a dog isn't feeling well, or hear a hummingbird chittering against the cacophony of background noises.

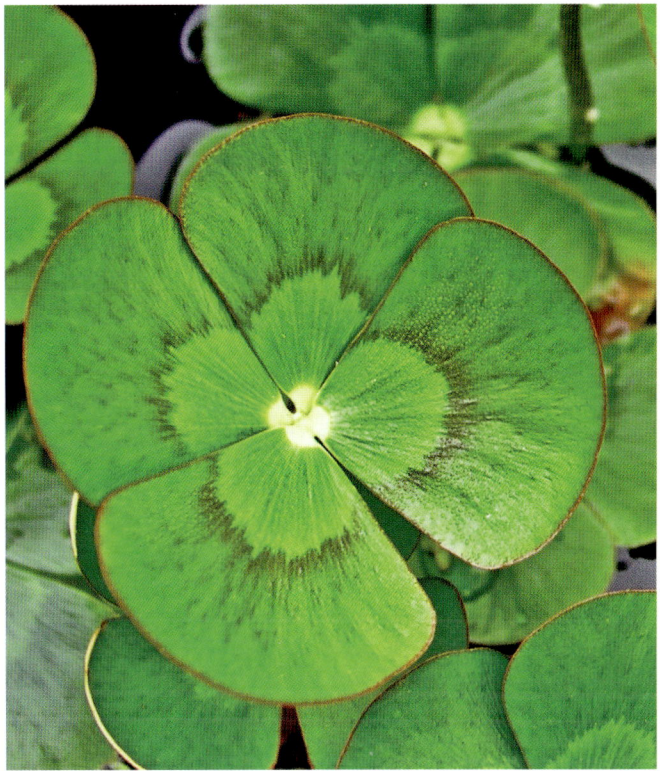

Most DIGrs have extra dollops of naturalist intelligence, coupled with the one called visual-spatial, which helps them see both the whole and the parts of their gardens—past, present, and future—in their mind's eye, and understand how everything works together and changes over time.

In gardening, it is often simply called "having a green thumb" which is mostly just paying attention to and acting on natural cues like when plants need watering, or when the soil is warm enough to plant seeds, when a melon feels ripe, or how the air smells before the approach of a summer rain.

What DIGrs Dig

Anyone with garden-woke eyes can easily spot a DIGr garden. They pop up in cities and villages, along country lanes, in upscale as well as less well-heeled neighborhoods, and in community gardens, and even perch on balconies of high-rise apartments.

While I resist comparing or romanticizing the ways gardeners do things, most DIGr gardens have several design and material elements in common:

- Seemingly random or higgledy-piggledy design as seem from the street; they are more often laid out from the gardener's perspective, from the house
- A fence or hedge with a gate or arbor entry, to set the garden distinctly apart from the larger, more judgmental world
- A clearly defined small lawn (if any grass at all)
- Meandering walks and paths, paved with bricks, flagstone, or broken concrete or just mulch or hard-packed dirt
- Walks and beds edged with stones, bricks, pottery shards, wood boards, or upended wine bottles
- Hodgepodge of plants of all types, sizes, and seasons, and an old queue of pots of stuff waiting to be planted (often in vain)
- Flowers, showy foliage, or painted objects to provide cheery color every week of the year, especially in midwinter
- Tolerance and appreciation of native or aggressive plants that more conventional gardeners call "weeds"
- Provisions for or benign acceptance of all sorts of wildlife, including both beneficial and a few not-so-welcome critters
- Containers stacked and hanging everywhere, of every imaginable material, including clay, wood, recycled metal, porcelain household fixtures, tires, and foot and head gear

---

DIGrs usually have front gates.

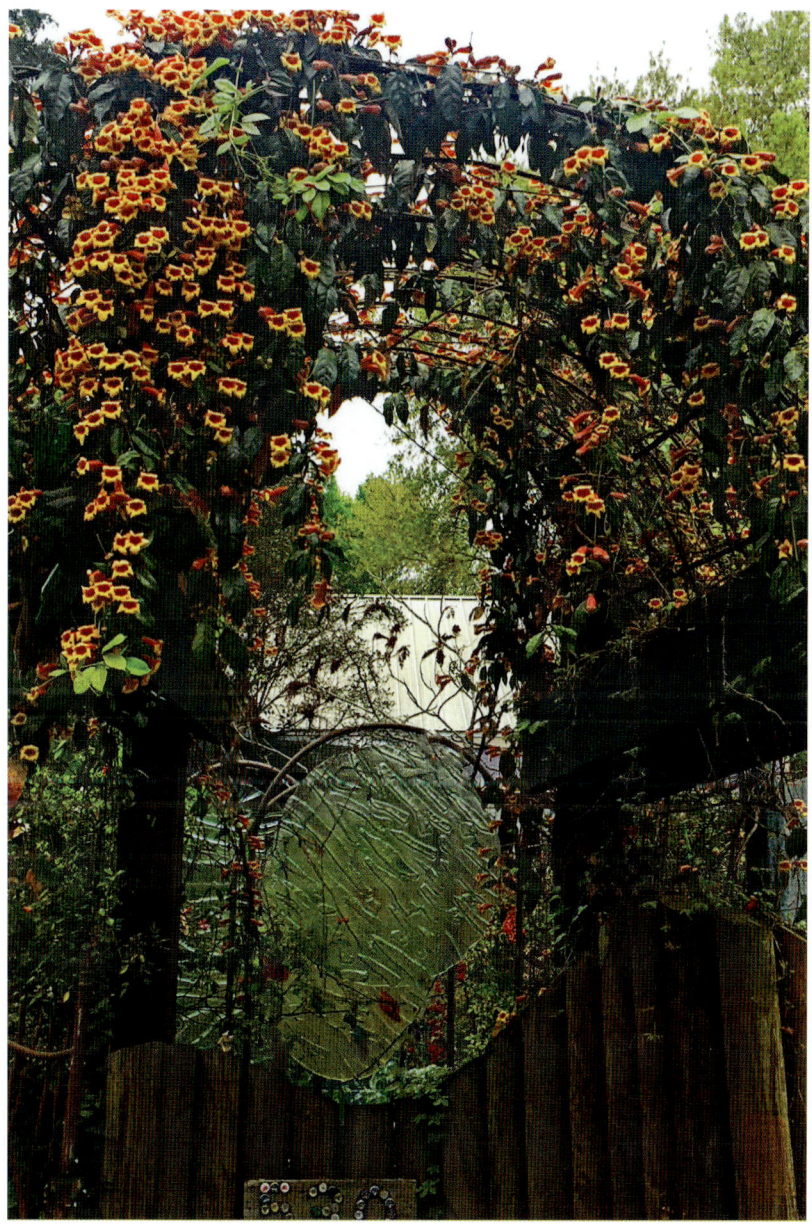

- Compost and mulch piles, open sacks of potting soil, old pots, and other "useful someday" stuff they can't bear to throw away
- Hoses, water buckets, and well-used tools and equipment kept openly close at hand
- Assorted garden art, both classic and whimsical, sometimes store-bought but often homemade from found objects

Note that these traits are by no means universal; and those gardeners with independent-minded approaches who land in upscale neighborhoods have to—by choice or neighborhood requirements—be more discreet, or hide their fervor in backyards.

Sadly, there is another near-universal trait: DIGrs, regardless of their social status, often have to deal with derisive mocking from less imaginative people.

You Might Be a DIGr If:

- You have local plant swaps circled on your calendar
- Look at your hands right now—got thorns, cuts, or dirty nails?
- You've rescued discarded plants or cuttings from neighbors' curbs
- You check your garden at night with a flashlight
- You have something planted in a milk jug
- You deal with weeds by just planting stuff that's taller
- A toolshed is a place of comfort—and you know where its spider hides
- You don't mind bad smells if you know what they are from
- You can amuse yourself with your thumb over the end of the hose
- You've found seeds in wadded bits of paper from the washing machine
- You can smell a good rain an hour before it gets to your garden
- All you want for your birthday is someone to haul mulch for you
- Finding plants on your porch left by a secret admirer makes your week
- USDA Plant Hardiness Zones mean ... nothing

Nurturing the Inner Child

Kids are okay with coloring exuberantly outside the lines; when do they learn to stop? Do we really have to get stuck somewhere along the line while growing up, including in the garden?

One of renowned child psychologist Jean Piaget's most widely accepted tenets is that children think completely differently than do older humans; they aren't limited by what they don't yet know.

He theorized that well-developed children need to "play" in order to discover new concepts and learn what works and what doesn't. As they begin to decide for themselves, they map out how they will approach things later.

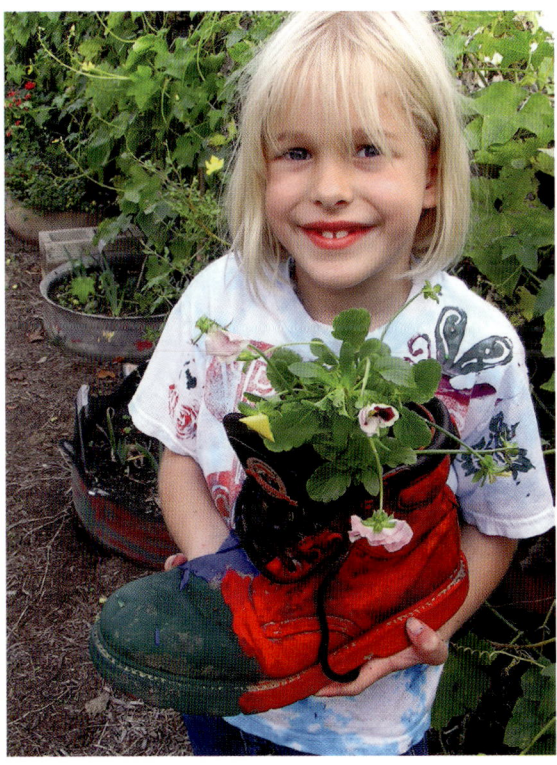

Children are encouraged to develop imaginations, starting with bright colors and moving objects, and trying to touch or taste everything. They eventually begin comparing stuff, learning how to manipulate and act on what they learn.

Older children more readily accept patterns for how things are done and follow rules (sports, music, dancing, behaving in public, etc.); this is when they start to conform with customs to "fit in" and get along with others.

As we grow up, along the way many of us settle into familiar routines. Based more or less on what and how we learned as children, we surround ourselves with likeminded others and accept those norms.

Sadly, by that point we often stop playacting easily and our joy of simple things becomes dulled.

DIGrs do not. Unlike status quo and formal gardeners who stick with the comfortable, and with what others believe, DIGrs continue to explore, imagine, pretend, create, build . . . to think outside the box. Holding themselves open to new ideas, they aren't afraid to try and learn from new things, and to even fail. And they adjust accordingly.

In short, *they play*.

And where better for adults, yearning for an earlier age of more color, senses, simplicity, freedom of worries and rules, to find that spirit again, than in the garden?

2

Gardener Coming Out

VISITOR: Hey, I just heard that . . .
ME: Don't interrupt me, I'm gardening.
VISITOR: Well, you were dancing around.
ME: Dancing is a *part* of gardening!

The slow-simmering realization that I'm a DIGr came more of a surprise to me than to my nearests and dearests.

Before I introduce a few classic independent gardeners, I should share a bit of what led to my personal quest to find out what motivates the countless seemingly blithe, free-spirited gardeners I've met, who somehow hold their heads up in a world that wants to mainstream them.

Along with plenty of forehead-slapping epiphanies that changed our courses, usually for the better, we've all gotten inspirations from others.

Of the latter, one of mine stands out in particular. The late Marc Cathey, president emeritus of the American Horticulture Society, loved to challenge small children, home gardeners, and fellow horticulturists with innocuous questions and thoughtful comments to provoke personal discoveries and a desire to learn more.

During a live radio broadcast with me back in the early 1980s he sat back in his studio seat and smiled, then turned me on my ear with a simple, conjuring question: "Can you remember the first time someone showed you a plant, or an insect, or anything else in the garden, and actually explained it to you?"

My immediate answer is lost now, because it took me a few years of digging deeply before I really understood his challenge.

Getting Started Early

My earliest plant memory, still vivid, is of when Mom's huge prickly pear cactus ate my brand-new inflatable beach ball. I was five, but I can still see in my mind Mom putting tiny bits of cellophane tape over the punctures.

Five years later, while looking for a school project plant, I bicycled out to meet Betty Pearson, the patient owner of my small town's garden center. I remember how amazed I was when, with a dismissive wave of her hand, she gifted a small potted succulent to me. It was portentous of things to come.

As a teenager delivering newspapers after school from my bicycle, I noticed how some gardens were distinctly out of synch with the majority—over the top with flowers, vines, structures, and quirky art. The rumor mill whispered about their creators.

Later, when working at Pearson's Nursery, I met some of them and discovered they were perfectly normal in most respects and quite interesting (though they didn't buy much).

I learned a lot more at that garden center, from plant production and landscape installation to dealing with fickle customers. Seeing a natural knack developing, a high school friend who had gone on to study horticulture in college convinced me to give that formal route a go.

The Four Women

Funny thing. My three siblings and I were raised in the starkly different gardens of our mother, two grandmothers, and horticulturist great-grandmother, and were exposed to similar opportunities and experiences. But though my sister and a brother eventually became serious garden dabblers, for the most part they were other-motivated, bent towards different routes.

Yet for some reason I was smitten early on and now can't be untangled from the sometimes-conflicting styles of those four women. I sussed out early that though they watched one another carefully, each basically gardened alone in her own world.

Mom was a self-trained, keenly observant nature lover who entertained everyone with the differences in bird feathers and how moths emerge from cocoons. She taught me to recognize seasonal bird calls, and which piece of straw was best for snagging doodlebugs out of their deep burrows. Wilma Gene's cherished potted plants had to be dragged in and out with every change of the season, and the chore always seem to come to me.

Her mother, Granny, was a simple, stoic country woman who exemplified gleaning simple pleasures from humble materials. She *didn't know nothin'* about gardening, and was okay with it. I occasionally helped weed her single bed of tall, butterfly-laden zinnias and cockscombs that reseeded every year.

Granny also tended a few potted plants, including a "mother-in-law-tongue" sansevieria stuffed into a recycled bucket that was gussied up with used silver kitchen foil. One of her most cherished possessions was an old concrete chicken she'd received as an anniversary gift from my granddad at a time when he couldn't afford better; it stood sentinel over winter daffodils until zinnia season rolled back around.

My other grandmother, Louise, was prim and proper, an unfailingly cheerful garden club stalwart with booklets filled with blue

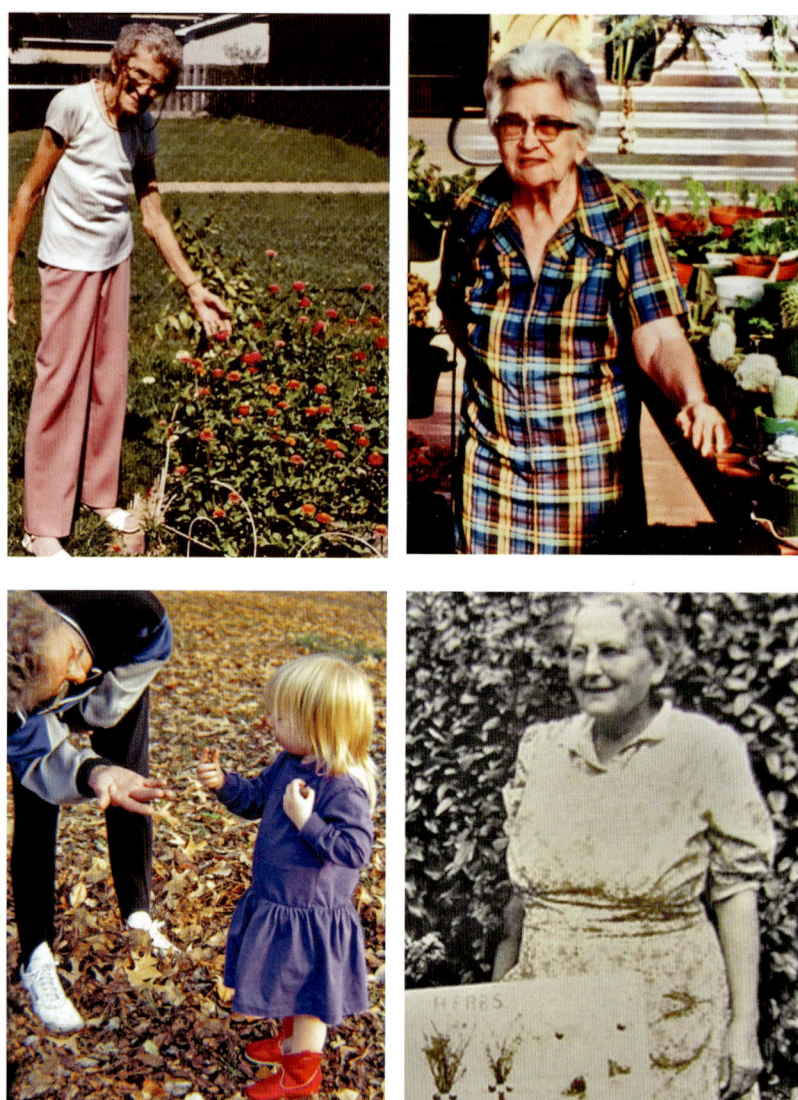

Plus Zoe makes five.

ribbons won for her hybridized daylilies, prized African violets, and beautiful flower arrangements; she taught me the value of planning, tending little details, and observing the esoteric rules for garden-circle social acceptance. She could be nice even when irritated.

My great-grandmother Pearl, who in her journal labeled me "Little Professor" when I was just ten years old, was a horticulturist, naturalist, flower show judge, and plant collector with row upon row of carefully labeled daffodils and chrysanthemums; she delighted in unusual edible plants including pawpaw, hardy citrus, and cassava. I still thumb through her meticulous journals of what flowered and when, and what her garden's wildlife was up to each month.

In spite of being somewhat crustily plainspoken, Pearl was very active in the garden club she helped form in the 1930s; however, as I gleaned later from furious notes in her old garden journals, she sometimes felt despondent when her circle of garden club friends scoffed at her passion for native songbirds and wildflowers.

There were others, of course. My grandfather showed me how to step on fallen pecans and feel through the bottoms of my shoes if they were firm and ready to pick up, or cracked from mold and not worth the effort. My dad explained the physics of how extra-long chains made his porch swing so much more languorous than Granny's frenetic short-chain swing.

But these four women, each strong in her way, taught me early on some important, if often clashing, philosophies that continue to guide my garden attitude all these decades later.

Opened Eyes

University studies all but spoiled gardening for me. I was snapped to attention by academia's hard scientific approach in which plants become crops to be programmed, managed, and marketed. One professor said that "A plant isn't worth a nickel until someone pays you a dime for it—what you don't sell, you gotta smell."

This was starkly different from what I was taught at the knees of the hands-on gardening women of my youth; their more laissez-faire approaches nagged at me through my entire career as a university Extension Service horticulturist. And over the years I found myself slowly shifting away from crop production consultant, towards being a more casual people-friendly advocate.

What people did—and how—started meaning less to me than why they did it. This, of course, led to criticism from my horticulturist peers that I was "dumbing down" our profession, which I learned to shrug off. Which led to more criticism.

Soon as I retired, I shed most of my horticulture mantle, striving now to be a more fulfilled gardener. I no longer worry about whether or not my tomatoes will ripen, because I know they can be fried while green, or made into green tomato chutney. Or just because my *simply planting* them gives me hope.

I became a remarkably mediocre gardener, and I hope to live longer because of it.

My Garden

They say you can tell a DIGr from down the street. Something stands out that says, "This person is a wee bit over the line."

And I'm right in there. Because of my horticultural background, my garden really should look better than it does, but it's a jumble of both common and unusual plants and overdone accessories, fenced so its general messiness is partially hidden from prying eyes.

I consider it a compliment when someone I meet for the first time realizes where I live and stammers something along the lines of "Oh—*that's your yard*?"

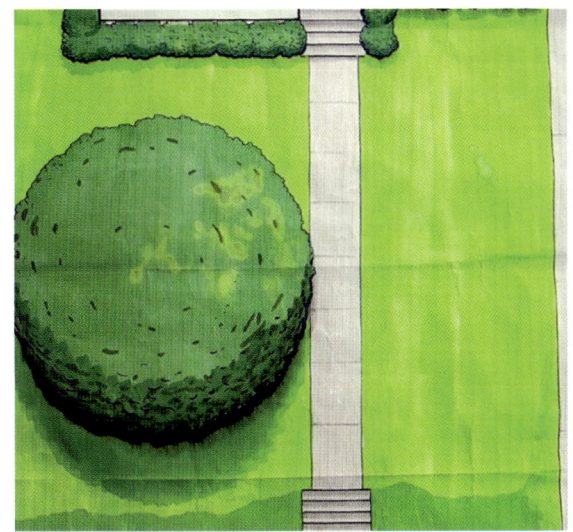

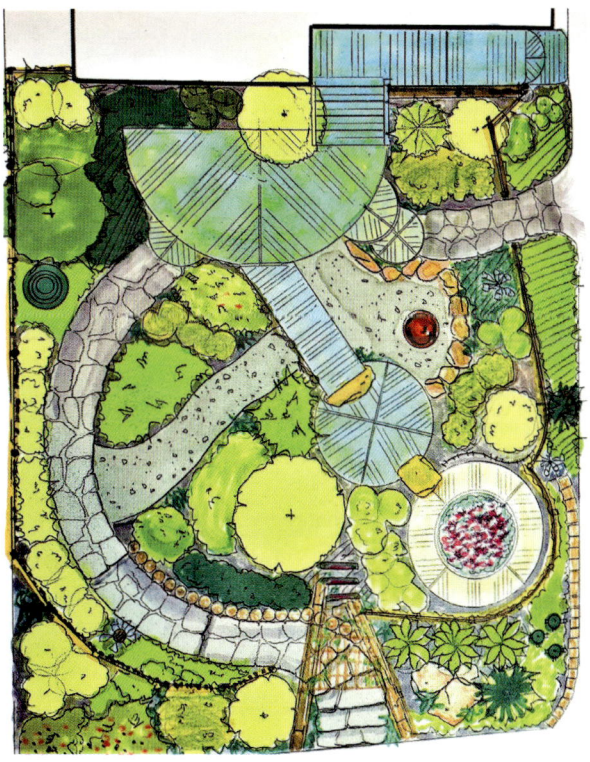

Before and after.

When my wife Terryl and I first moved into the bungalow on a clay hill where we raised our children, I was sure from the start that the wall-to-wall St. Augustine grass and overgrown shrubs hugging the foundation of the house simply wouldn't do.

It was distinctly bad *feng shui*, with steep steps from the street to the long, straight "poison arrow" of a sidewalk cleaving through a soulless carpet of lawn and leading to yet more steps into the house. Coupling my dislike of mowing grass in general with an already well-developed love of plants, meant the yard needed a complete makeover.

So, in a move that proved, er, unsettling to the neighbors, I killed all the grass. On purpose. The yard became a *de facto* bare garden slate, ready to jump in any direction. I had ideas, of course, but knew enough to get a little informal advice from landscape architect friends.

Rejecting cookie-cutter suggestions from those who were playing it safe, I landed with Rick Griffin, a creative landscape architect friend who nailed it for me when he said, "There are too many lines already in your life, Felder. *Let's do some circles!*"

We reversed the garden so it faced the house, not the street. "You don't owe anything to strangers zipping up and down the street," he said. "Let's make it a private garden to view from your front porch."

All this set the garden up for decades of playing, steadily planting or swapping out new treasures and tweaking whatever needed it.

Kaleidoscope Having a Stroke

Vertical is important to me. Scattered throughout are vine-covered arbors and trellises, some sporting collectable artisan birdhouses, the most strategic, lit up with strands of small green or blue lights or hung with Spanish moss.

A startling scattering of glass bottle trees festooned with fairy lights prompted my mother to declare that the yard "looks like a

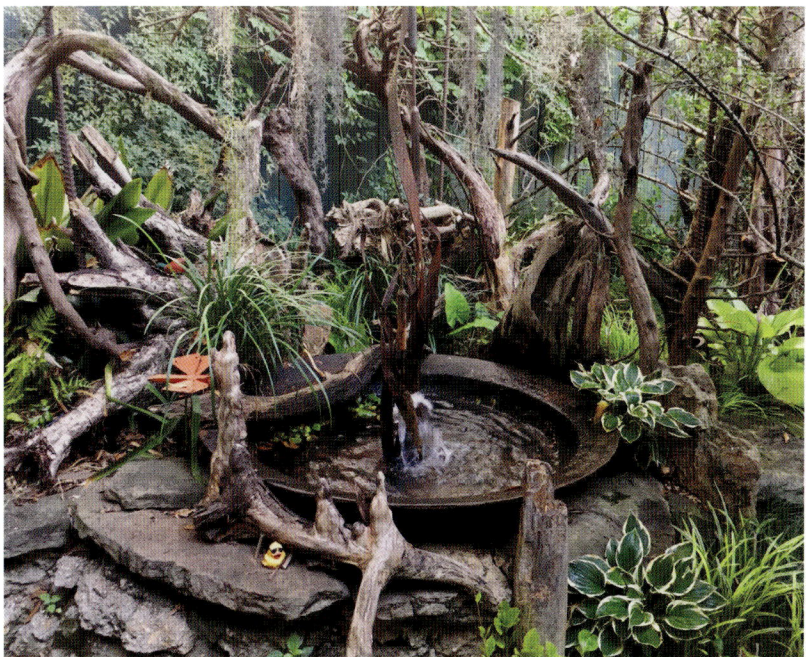

kaleidoscope having a stroke." But I dearly love how the sun shines through the colorful glass, how they reflect and transform neighborhood lights at night, and how icicles and sometimes small piles of snow settle on them in the winter.

A very shaded corner is now a "stumpery"—a collection of large, weathered tree stumps and logs covered with mosses and lichens and planted with ferns, hosta, iris, liriope, heuchera, and other shade lovers.

As for sculptures and other artsy accessories, here are just a few of my most treasured: flower beds edged with upended wine bottles, a working chandelier, an authentic copper moonshine still, a colorful glass birdbath, metal agaves spray-painted realistically, a flock of pink flamingos (one signed by Don Featherstone himself, the man who patented them in 1956), a six-foot graduated stack of

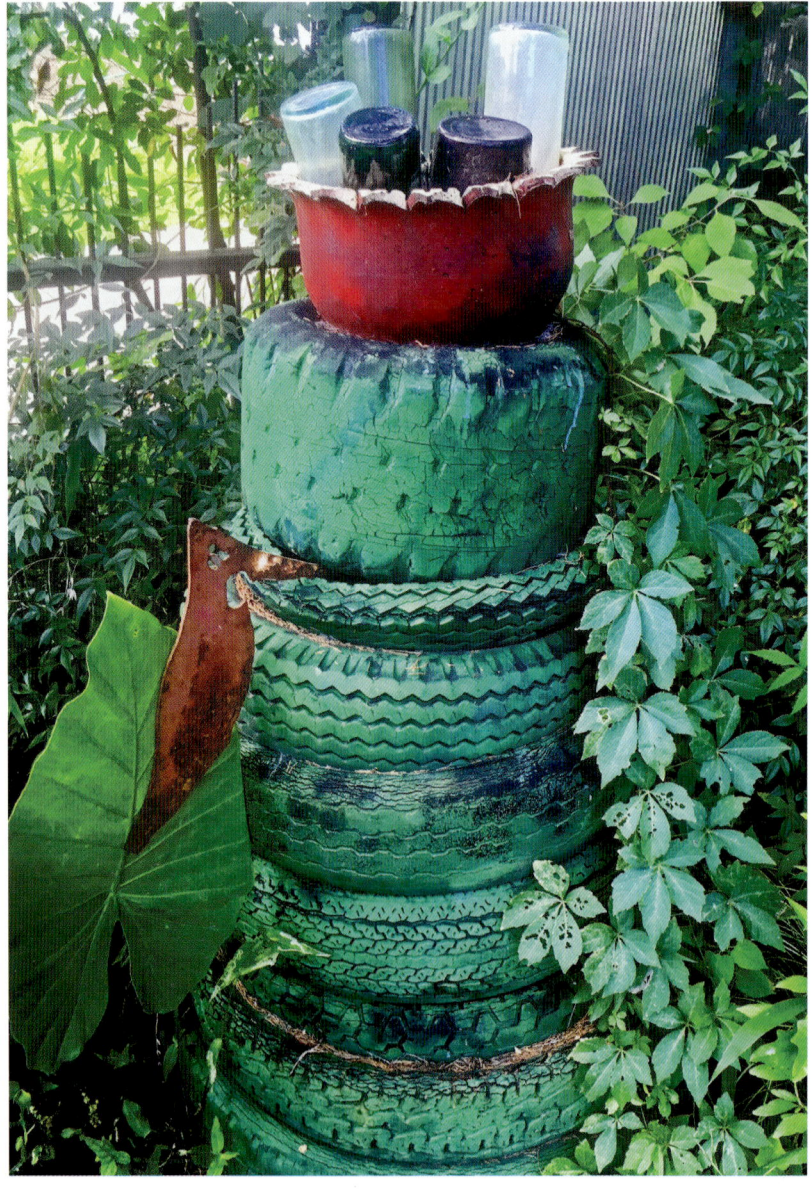

tires painted green like a shrub, a five-foot metal bird sculpture that doubles as a beer bottle opener, Granny's prized concrete chicken, a trio of salvaged iron fire hydrants painted teal, numerous flowerpots made of cut and inverted car tires, and a life-size female mannequin draped demurely with a cape lent by a modesty-minded friend . . . and more.

So, sure. Over-accessorizing is not an easy concept for me.

There's ample evidence of garden eccentricities I take for granted but which have been pointed out by bemused visitors. Most fellow DIGrs will recognize some of these in themselves:

- I love just . . . digging. Will turn a patch of dirt over and over, sometimes for weeks, until I reluctantly concede it's ready to plant.
- Windowsills are lined with water-filled jars of rooting cuttings.
- A big rosemary shrub started out as rescued bit of mashed potato garnish at a restaurant that I saved and rooted.
- There's that scraggly row of potted plants on the drive, hoping to be planted some day.
- Visitors occasionally leave odd new plants with little notes on the porch.
- An oversized, overloaded plant stand has to be unloaded and wrangled indoors and out every autumn and spring.
- Tree branches are hung to keep birds from flying into the large mirrors I mounted on walls to make the garden seem bigger.
- Shards of broken art glass top a back fence as a colorful burglar deterrent.
- The potting bench on my porch is an old porcelain sink framed with a paneless wooden window for a feeling of enclosure.
- There's a Wildlife Habitat sign by the street to let neighbors know I'm doing it all on purpose.

What started out as wall-to-wall mowed lawn and a straight sidewalk is now an informal, low-maintenance horticultural seventh heaven. It's my refuge, my playground, my experimental plot.

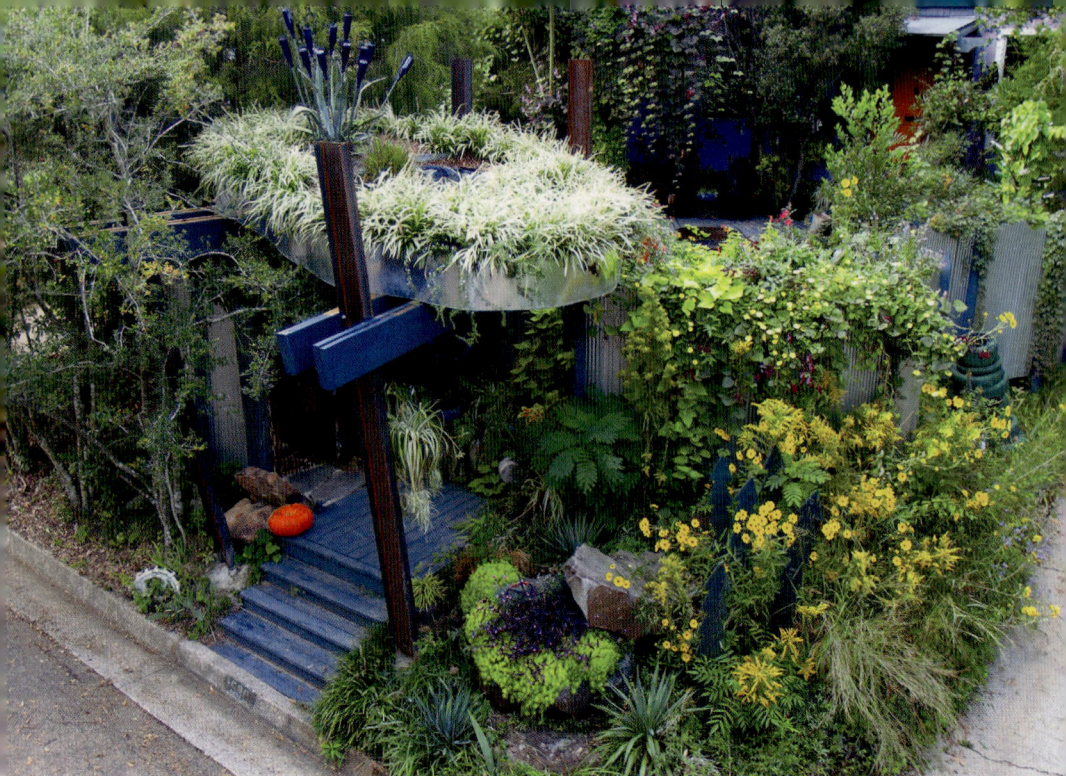

Birdseye of Felder's front garden.

And it's been celebrated just enough for the neighbors to sorta take pride in it. A new neighbor said that one of her house's selling points was that "it's on Felder's street—*but not too close.*"

It's an exploration guided by good people sharing passalong plants and the stories and connections they all have in tow.

Felder Truck

You've probably seen an old farm wagon planted with flowers. How about an entire truck—that not only runs, but is driven countless thousands of miles a year—filled with plants?

Country Girls and Other Hardy Mums

I'm a plant collector nerd, with dozens of rare sansevierias, heirloom narcissus, and other bulbs. But I'm especially obsessed with a special type of hardy "heritage" garden mum named *Chrysanthemum* x *rubellum*.

The most common cultivar of this autumn mainstay across the hot, humid Southeast is called "country girls"—a folk name for the 'Clara Curtis' cultivar. The pale pink flower was a favorite of both my great-grandmother and a gardener named Dirt (more on him later), but I have seen it from Texas and Oklahoma to South Carolina, Wales to Japan, and even Canada where temperatures dip to 30 below zero.

In addition to the common pink one I now have a couple dozen others, including red, white, yellow, cream, and bronze. And I am getting cuttings (which root in just three weeks) into the hands of nurserymen in an attempt to get them into the popular mix.

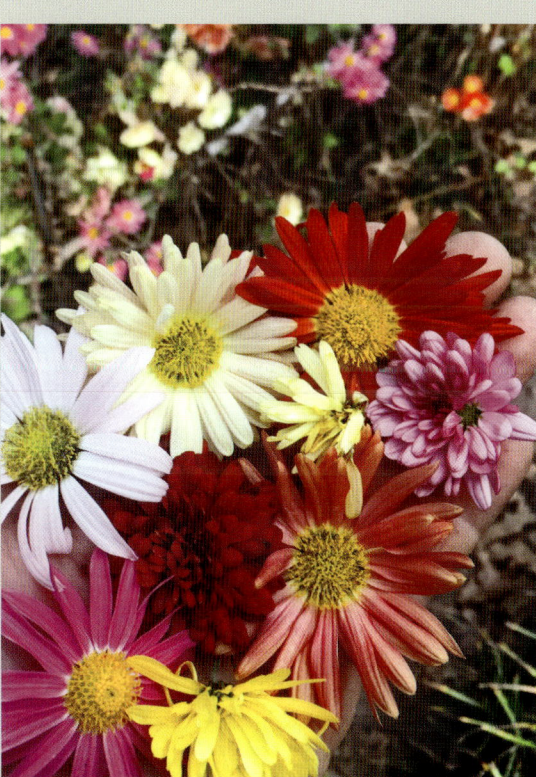

So, I'm moving ahead, looking back to what used to work and still does.

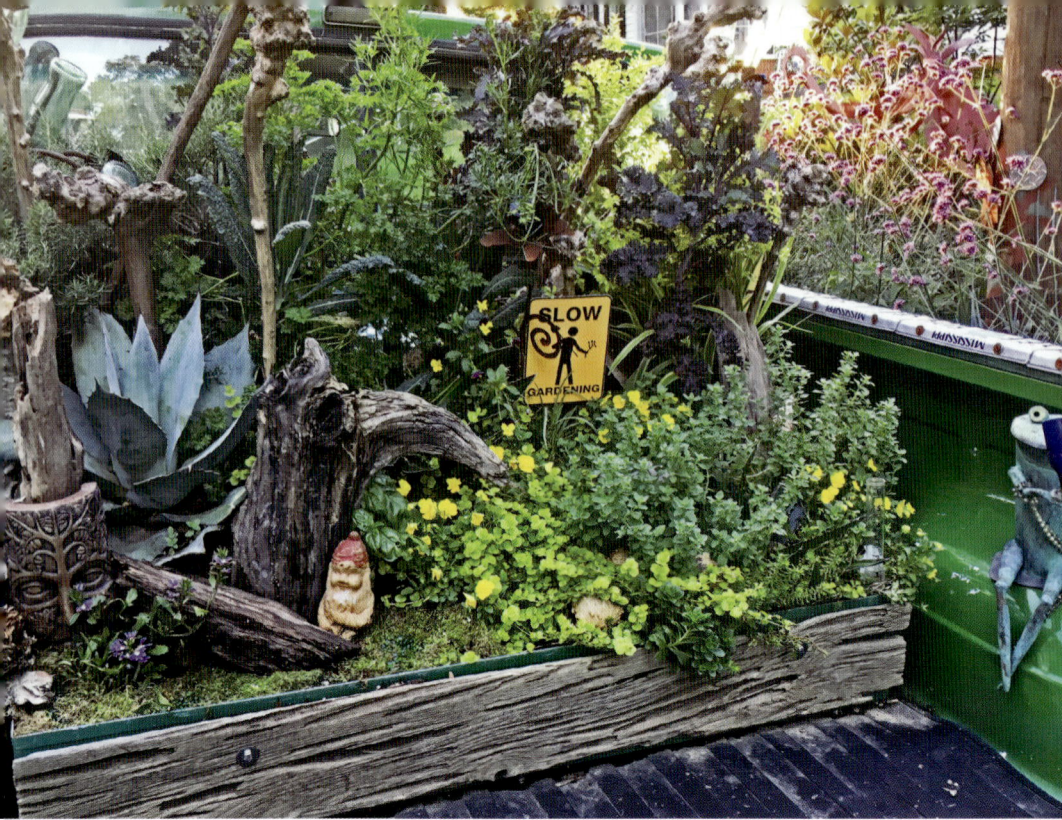

Some thirty-odd years ago, in response to someone moaning about not having a place to garden, I tasked myself with developing a simple garden in the most difficult place ever. Needed to be inexpensive, alluring, edible, and low maintenance.

To up the interest and challenge, I settled on creating a complete garden in an extremely unlikely place: In the back of my hard-working '87 F-150 pickup truck.

I started out simply, nestling a bag of potting soil in the windless eddy against the back of the cab and tucking in a handful of flowers and vegetables that got watered maybe once a week.

Couple of years later the short-term sacks of potting soil were replaced with a custom-designed-to-fit metal box filled with lightweight potting soil. I planted small shrubs, herbs, vegetables, and

Felder Fesses Up

Some garden experts opine that it's best to always be positive. But as much as I'd like to gloss over how gardening isn't always rosy, I'd rather maintain credibility with folks who know better.

So, for over four decades of writing weekly newspaper columns, I end every year with a confession of my garden foibles and failures. I've had my share.

Countless plants have withered from poor soil preparation, neglect, or pest problems beyond my ability to control. Grasshoppers, slugs, squirrels *et al* have eaten their share and more. Plants have been stolen, and there are those withering plants in my "plant purgatory" queue looking silently reproachful while waiting to be rehomed.

I have dug water gardens, and filled them in. Rearranged or removed arbors, walks, and decks that didn't work well. Accidentally squashed good garden critters while moving stones, and found dried-up lizards in my plant room window.

My eyes have swollen shut from poison ivy, hay fever has driven me indoors, and I squeal like a little girl if I run into a spider web. I have offended neighbors with inappropriate yard art.

Oh, *there's more*. But you get the idea. Better next year, eh?

perennials, and if something didn't make it, I simply yanked it out and stuffed something else in the hole.

After over thirty years and three hundred thousand-plus miles of trial and error, I've found a surprising number of small, compact plants that survive summer heat, drought, winter freezes, and occasional blusters of crosswind when I pass big trucks. Now about all I do is simply swap out summer and winter annuals and pull weeds as needed.

Oh, and I over-enhance my garden with accessories including bottle tree sconces, birdhouses, rain gauge, a copper frog, and gnomes.

Point is, if I can garden all year in the back of a pickup truck, nearly anyone, regardless of skills or confidence, can do it on a patio or porch. With guidance even little kids can do it.

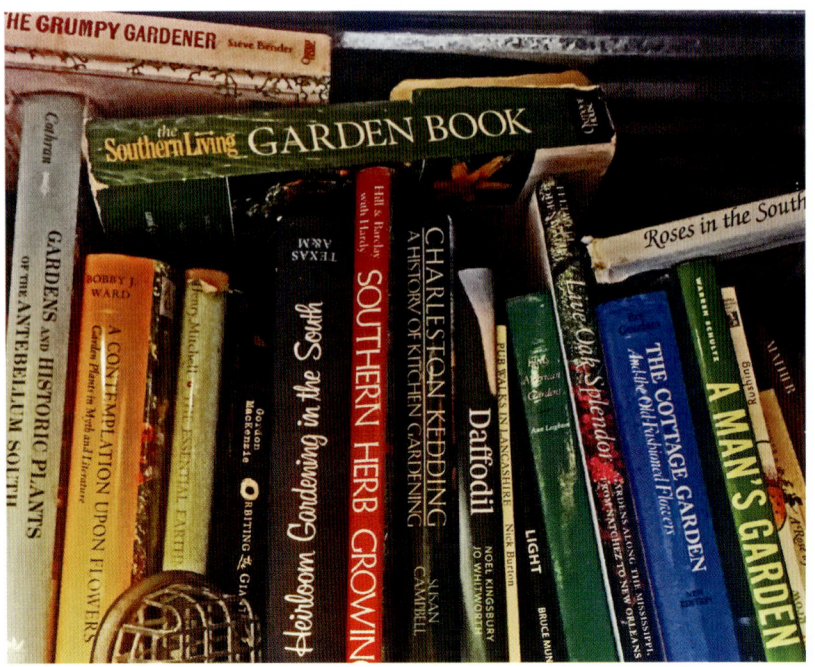

Where do old books go?

Moving It On Down the Line

Like playing with and shepherding my children through their early lives, guiding my garden has been a maturing process for myself as well.

However—and this may seem a bit macabre—like most gardeners, as I play in my little corner of the world I sometimes ponder its ever-changing ephemeral nature. And, back of my mind, I am aware that sooner or later my garden will pass on to someone else's care—and they're not gonna like some of it.

I've seen this dust-to-dust happen too many times, how very few gardens outlive their gardeners, at least for long. And when a gardener is laid into the Great Compost Pile, the inevitable question quickly arises: What happens to what's left behind?

While a few plants are left to fend for themselves, family and friends usually divvy up the best and the rest are regretfully cut down or dug up and discarded, and someone has to deal with the well-worn tools and groaning shelves of lovingly earmarked gardening books also left behind.

And by the way, under which rose bush did I tell anyone my old dog was buried?

There's a point to all this. It's one thing to "gather ye rosebuds while ye may . . ." But it's also a good idea to share what you can, while you can; unlike a tattered old photo or a jpeg floating, lost, in an internet cloud, plants and laughter can live on. Especially if a DIGr was involved.

In the next few chapters, I hope to inspire you with stories of others who feel the same.

A Man Called Dirt

> Everything starts from dirt, everything returns to dirt.
> —Leon "Dr. Dirt" Goldsberry

Of all the men and women I have met over the years who held their heads high while "doing their own thing" in their personal gardens, the most conspicuous of them all, hands-down, was a man who called himself Dirt.

His celebrated Southern cottage garden, now long gone, was an astonishing repository of heirloom and garden-worthy native plants, and the man himself changed the way people in his region saw their gardens—and their roles in the gardening community.

I met him largely through chance. My gardening friend Rita Hall, an English gardener who had landed in Mississippi, had challenged me to show her some authentic Southern cottage gardens. We spent a day or two every season in my old pickup truck, cruising the outskirts of small towns and along country lanes looking to find those exuberant "folk gardeners" whose yards overflow with classic passalong plants and the lore that trails along.

One morning back in the spring of 1997 we lucked up. We were talking over the fence with an older woman in the small village of Edwards about a glowing stand of newly sprouted burgundy cannas

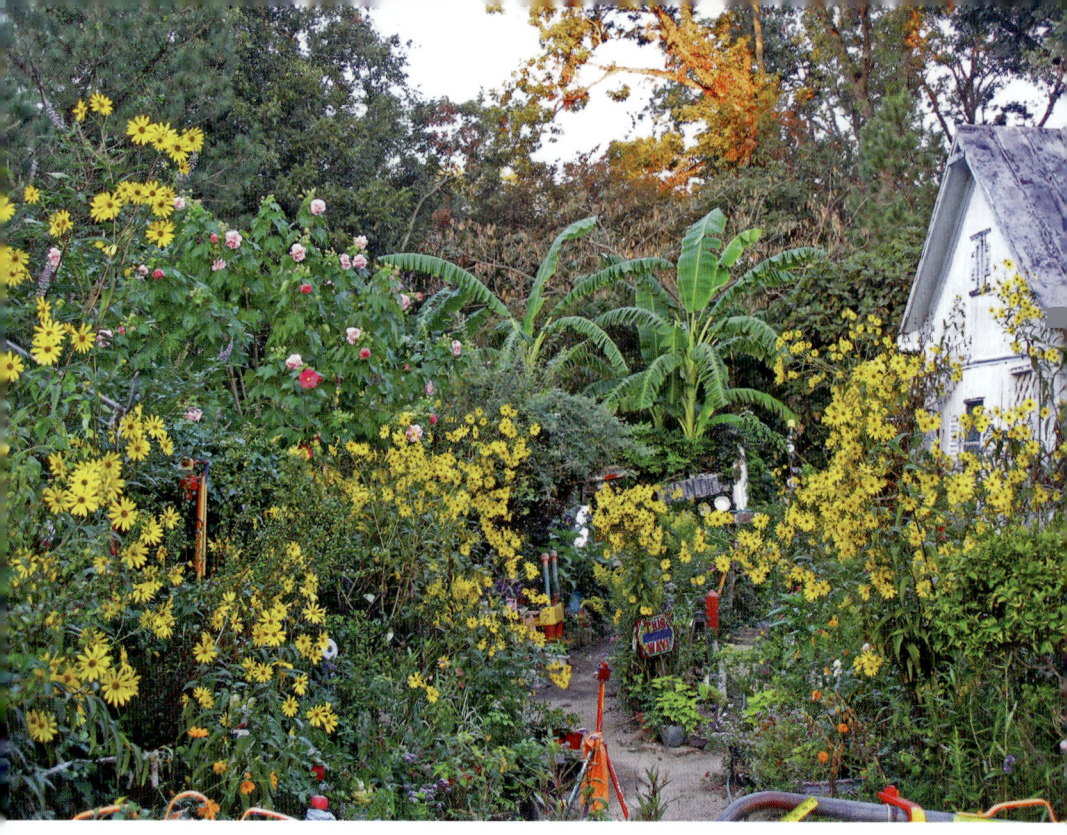

when she casually said, "If you think this yard is something, you ought to go see my nephew's around the corner."

When Rita and I headed the few blocks over, we were stopped dead in our tracks by a fenced garden crowded against a busy railroad crossing with passing trains roaring by every few minutes. The overstuffed "total yard show" was bursting its seams with sprawling and clambering plants festooned gaily with colorful hand-painted signs and slogans.

It took a few moments to take it all in, to absorb what we could. Because the garden was laid out in the classic backwards-facing cottage style best viewed from the house rather than the street, to a casual observer it looked like a jungle, with only a narrow, hard-packed dirt path leading the eye just a few tantalizing feet before disappearing into the vegetative smorgasbord.

Just Call Me Dirt

For a few minutes, even after calling out several times, there was no sign of the gardener. But just as we started to walk back to the truck, a man revealed himself to have been silently standing nearby the whole time, hidden in the flowery overgrowth.

He was a striking figure, six and a half feet tall with broad shoulders, a colorful du-rag bandana wrapped over his short-cropped hair, and a sweat-soiled towel draped around his neck. Looking all the world like a swashbuckling extra out of a pirate movie, he leaned warily on a long-handled brush cutter called a kaiser blade with its large, curved steel knife used to chop woody weeds and errant snakes, and just glared at us.

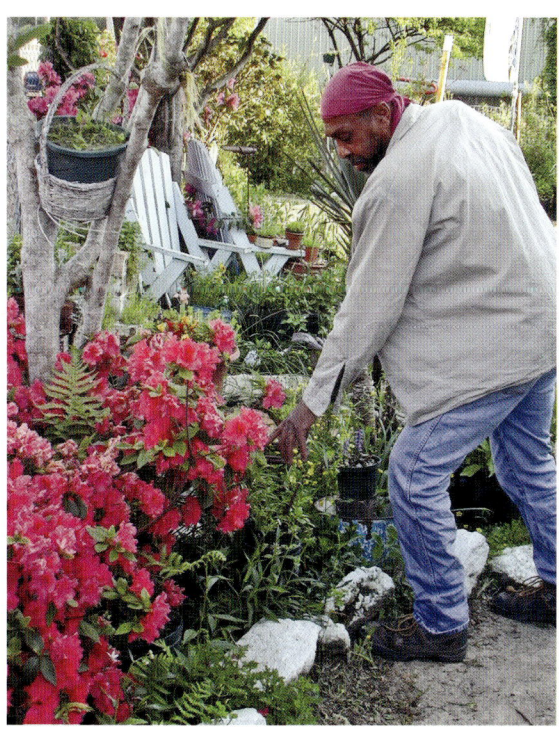

When Rita and I started to gush over his incredible assortment of plants—only a small portion of which we could see from his gravel pull-off in the street—he simply said, "It ain't my garden." No further comment, just a suspicious, almost judgmental stare.

We were persistent but got an emphatic repeat of *"It. Ain't. My. Garden."*

Even after identifying ourselves as just gardeners looking for other gardeners, which seemingly made no impression, all we got in response was "Just call me Dirt."

And with a dismissive shrug he saw us off.

Opening Up

I came back, dropping by two or three times over the next few weeks, always with a passalong plant to share (Dirt could never resist a plant from a likeminded other). Slowly, after realizing that I wasn't just another pesky plant tourist, Dirt opened up. He eventually admitted that he knew me all along from listening for years to my radio garden program.

With time, the guarded, nearly reclusive man formally named Leon Goldsberry revealed his gregarious hunger for sharing plants and insights with trusted fellow flower children.

He and his ninety-year-old mother Millie, who occasionally ventured outside to chat, were living in her father's century-old ramshackle tin-roof cottage painted white trimmed with pale "haint blue" based on a widely held superstition of being able to thwart evil spirits.

He often regaled me and other visitors about Millie raising him in the garden during a difficult childhood. Being a too-tall, sensitive youth who brought handpicked bouquets of daffodils and wildflowers to teachers, he was taunted cruelly by others, which only drove him to spend more time puttering around the confines of the fenced garden.

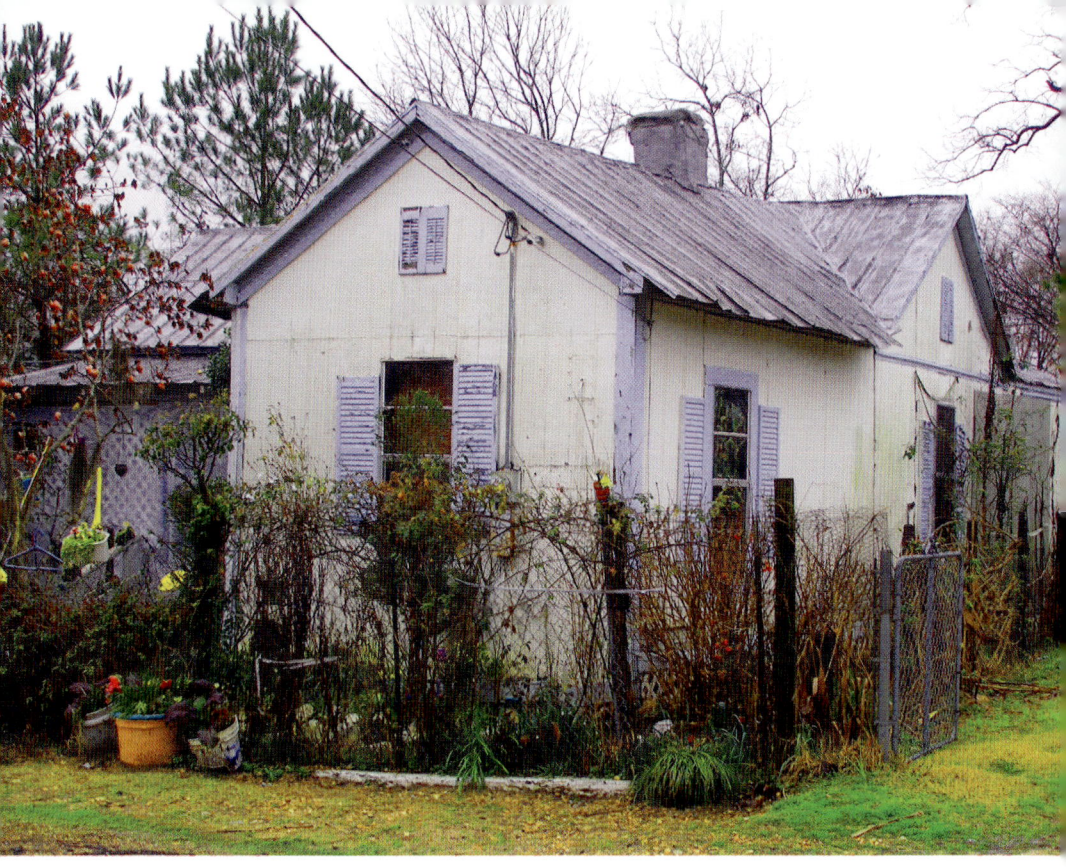

After a brief stint at Rust College, he avoided the wartime military draft by becoming an expatriate, working for nearly thirty years in Canada before returning home to care for his aged mother. He found to his dismay that he had changed more than those who stayed behind in the small Mississippi hamlet, but tending Millie's garden helped.

For all the years I knew him he steadfastly insisted, to anyone who gushed over his floral bowery, that the garden wasn't his. "It's part of my faith that created this. The flowers aren't mine; I just take care of them.

"I consider the garden—my mother Millie's garden—a folk garden, because of its old-fashioned plants, and all of the 'art' made of things other people threw away that I work in artistically with the flowers."

Whence the Name

When pressed, Leon Goldsberry would grudgingly tell his version of why he chose the Dirt moniker.

"Truth is, I don't know my African ancestors' real names, just the slave name we were given. And get real—*do I look like a Goldsberry to you*?

"I used to listen to garden experts on their radio shows, and loved the names they gave themselves. Dirt Queen, Dirt Doctor . . . I loved it.

"So now I just go by Dirt. Dr. Dirt."

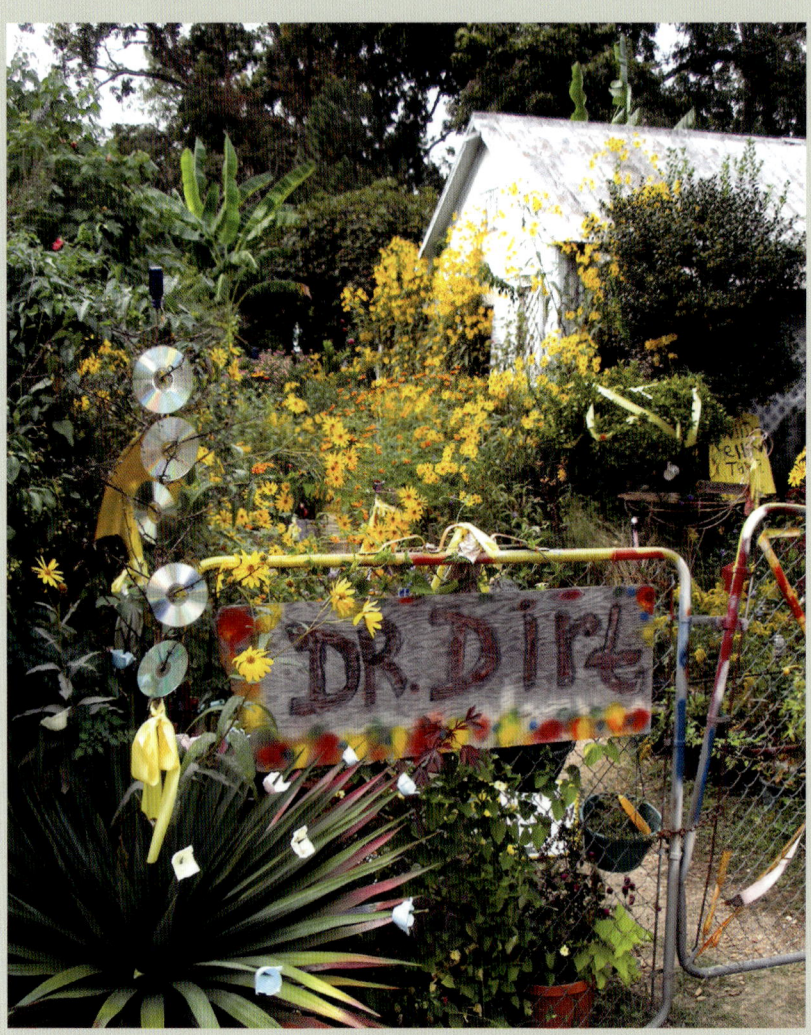

Dirt's Garden

The garden itself was in a low, flood-prone bottomland that drained into a ditch alongside the railroad. Its soil was a heavy but well-drained silty loam that retained nutrients. Dirt meticulously raised his flowerbeds using soil dug from trenches and channels which helped excess water drain away during wet spells. He often set potted plants in the ditches to help soak up water.

In the center of the property Dirt created a semi-hidden inner den, a private sanctum with walls and partial roof cobbled out of roofing tin, old boards, doors, and other hodgepodge materials, including

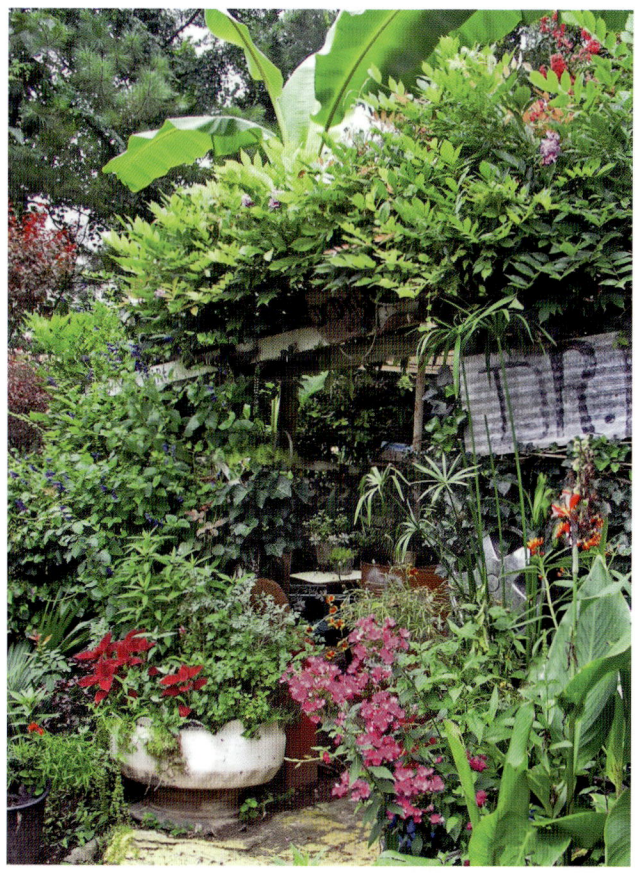

colorfully painted concrete flooring partly covered with outdoor carpeting. He painted his salvaged outdoor furniture and containers a cheery bright yellow to match his favorite flowers.

The lair was expertly camouflaged behind towering clumps of banana plants and a huge, tangled wisteria vine, and was used for refuge from summer sun and winter's cold winds, and cooking over an outdoor fire.

Many garden rooms were accessed through a maze of sometimes dead-end paths that wound throughout the property. Some planting areas were shaded by ancient trees and towering shrubs that protected flowers in need of respite from the harsh Mississippi sun; others were sundrenched and hosted homegrown vegetables and culinary herbs every month of the year. All were hit-or-miss combinations of plants, with the soil underneath carefully kept weed-free in a "dust mulch" of dry clay churned regularly with his sharp hoe.

He judiciously metered out rainwater collected in buckets and carefully crafted water-retention ditches, used mostly homemade compost for fertilizer, and never used pesticides of any sort.

Dirt's Plants

No question, Dirt was a plant hoarder, and Millie's garden was his Oparian treasure trove of mismatched beauties. When visitors came by, they were, as he put it, "blown away in a good way, and impressed by how many of these plants they remember from their parents and grandparents, but are hard to find in now in garden centers."

He threw himself into growing everything he could find, each an original family heirloom or native wildflower rescued from the edges of fields and roadside ditches.

Whatever would grow in plain unimproved soil, or in anything that would hold potting soil, he'd give a go, improvising what he needed on a limited budget.

It wasn't long before he moved beyond the astounding assemblage of plants that were already growing in the three-generation homestead—including a hydrangea planted by his great-grandfather, its deep purplish-blue flowers indicating the acidic pH of his soil—and started gleaning precious plants from nearby gardeners.

"I like plants that you get from other folks," Dirt explained. "I'm the type who goes to someone who also grows plants, who likes to share and swap." Some of his acquisitions, lined up in scraggly queues, waited so long for a garden opening, their labels had faded; sometimes the pots were there so long the plants themselves had disappeared.

Always on the lookout for new starts of plants, Dirt tried to propagate everything he came across, gathering seeds and plastic bags of cuttings wherever he went. Those that rooted or divided easily or came up from seed quickly were shared with visitors.

And he was always prepared. Throughout the year his small kitchen table was cluttered with boxes, bottles, and squares of newspapers piled with collected seeds, and jars of water with rooting stems of cuttings. For that matter, his home's walls were papered with pictures cut from garden magazines, to inspire him during the dreary wintry days and nights.

In his garden, rooting beds were stuffed through all seasons with cuttings of roses and other shrubs and perennials; more cuttings and seedlings crowded pots filled with plain unamended garden soil (which was one of many ways he flaunted more formal horticultural protocol).

Inventory

While closely reviewing my hundreds of photos taken through every season over many years of Dirt's garden and his weekly flower arrangements, I uncovered and verified many of the otherwise-forgotten denizens of his botanic wonderland. Without even counting the several varieties and cultivars of many, I have documented 193 different species of plants; I'm sure there were others.

And this doesn't begin to touch on the artificial flowers and foliage he cheerfully and effectively used to enhance his natural bounty.

I will take you through the highlights here, but a full list of Dirt's astounding array of plants is at the end of this chapter. The names of the plants are primarily what Dirt called them; many have several different common or folk names, so I have included alternate and Latin names for a few of the more obscure or unfamiliar ones.

Trees

Long before he moved back and started gardening anew, Dirt's three-generation family homestead was shaded in the back with an old pecan tree, a couple of oaks, and a cypress. He quickly added others, which in addition to having showy seasonal flowers and fruits, framed his garden on the sides and shaded it in the summer while providing leaf litter for his compost.

Seedlings from his Chinese parasol and offsets from the colony of wild Chickasaw plum were easy to pot up and share, and he rooted many cuttings from his summer-flowering vitex with its bright streaks of blue spikes.

Shrubs

The mainstay of all gardens is shrubbery, and, after years of carefully noting what flowered around town, Dirt managed to end up with something in bloom, from early spring quince to late winter camellias, every single day of the year, regardless of weather.

He often used their flowers, seedpods, and foliage in seasonal bouquets, and welcomed their colors, fragrance, and pollinators all year (because of his mild climate, he had active honeybees and small butterflies in his garden even on sunny days in the dead of winter).

And he was masterful at acquiring them by hook or crook. "You need to check my purse on the way out your garden," he would joke, "because you KNOW I got me some cuttings!"

Roses

Dirt loved roses, especially the ones he grew from cuttings taken in area cemeteries and while on visits to other gardens. Visitors especially admired the big everblooming 'Paul's Scarlet Climber' clambering up one end of his cabin.

One of our proudest projects was the planting of heirloom roses and bulbs at the early 1800s "garden park"–style Greenwood Cemetery in Jackson, Mississippi. Enlisting help from rose-loving volunteers, Master Gardeners, and the Old Garden Rose Society, we ended up planting over two hundred shrubs.

We chose plants that were proven to be hardy when grown on their own roots (not grafted) in unimproved graveyard dirt, and those that were repeat bloomers and disease resistant. Some came as rooted cuttings from our own gardens, but when Dirt and I lectured at the Antique Rose Emporium of Texas, we asked for roses in lieu of an honorarium; we handpicked all my truck could carry—a quarter ton of them—and hauled them home.

NOTE: Though different roses will perform better in other parts of the country, for our hot, dry, mild-winter cemetery we have pretty well stuck with the ones best adapted for our climate including carefully selected Old Garden Roses, dependable modern everblooming polyanthas, and floribundas.

Now the plants are being spread to other gardeners by volunteer teachers who hold annual pruning and rooting demonstrations right in the cemetery.

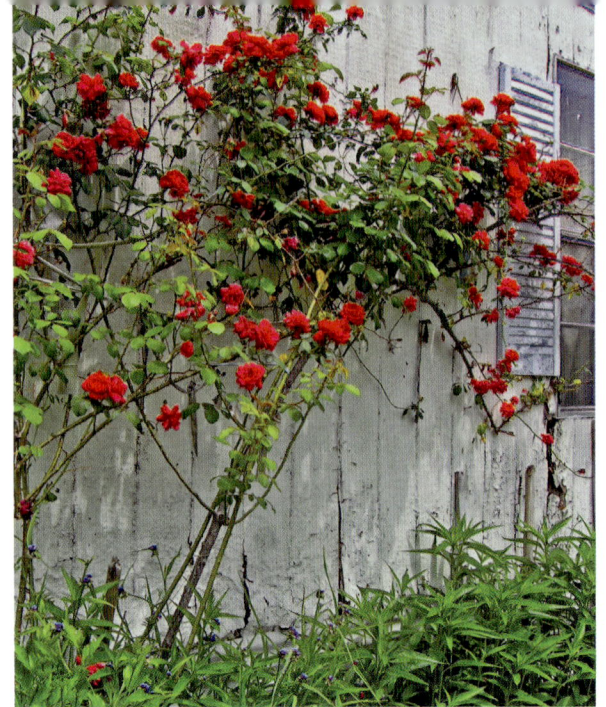

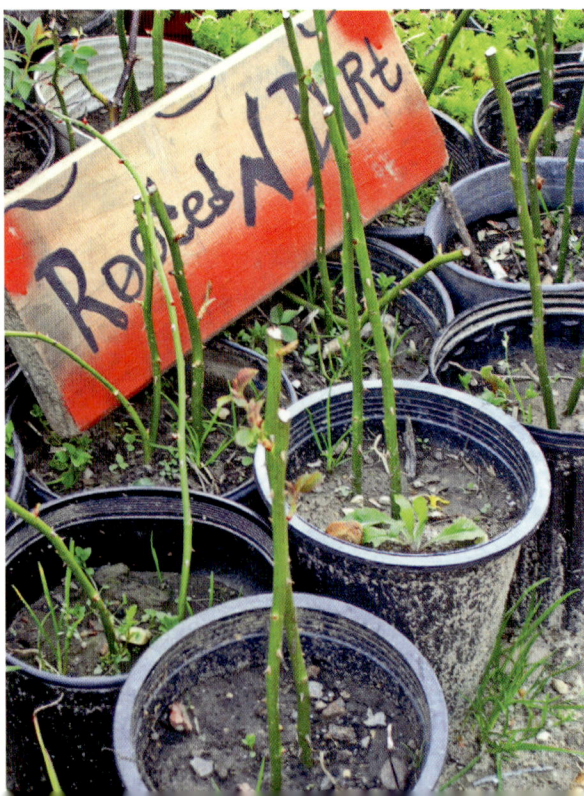

Dirt pruned his roses whenever he needed cuttings, which, against all conventional wisdom, he stuck into pots of just plain dirt until they rooted. No rooting hormones, no covering with plastic pots or tents, just water and stripping the leaves. And for the most part, it worked just fine (though to be honest I think it was partly because he figured out what roses rooted the easiest).

Vines

Prominent features of Dirt's garden were rattletrap arches made from wayward tree branches and cast-off materials, mostly bent metal pipe, upended broken garden tools, and PVC. All were painted in whatever color he had at the moment, sometimes repainted with whatever suited his seasonal mood, and often as not dripping with Mardi Gras beads.

And all were coiled or draped with vines, both leafy and flowering, sometimes several overlapping. If something got in his way, he simply wrapped it around itself, rarely pruning something for sake of size. "Mississippi is a jungle," he would proclaim as he brushed underneath the more unwieldy ones. "It's supposed to look and feel like one."

One of his favorites was the rambling pink-flowered perennial vine called rose of Montana (*Antigonon*), which he called Miz Floyd's coral vine after the generous neighbor down the road who first shared it with him.

Herbaceous Perennials

Perennial plants that came back year after year were often shared as cuttings or from divisions of clumps. Most were planted in plain dirt, sometimes with a little compost mixed in to help get them started. Every year or two they might get a hit-or-miss scattering of all-purpose garden fertilizer, if they got any at all.

Being a lover of tall stuff, Dirt often left faded or frost-killed flower and seed stalks alone, seeing them as having a beauty of their own. Many were spray-painted for extra color in his garden, especially in the winter.

He treasured the lush banana plants that shaded his hideaway "office" but the perennials that made him beam the most were the tall narrow leaf sunflowers and goldenrods that lit up his autumn garden.

When sharing extra plants, he usually mentioned how they wouldn't care if the receiver was black or white. "Plants are color blind, they don't care who your mama 'n them are."

Annuals

Dirt had plenty of annuals that sprouted every spring or fall from seed dropped to the bare ground, but he usually collected a few seeds of each in little piles, envelopes, and jars on his kitchen table. Some, like zinnias, celosias, and peppers dried on their own; seeds of tomatoes and other moist fruits had to have their seeds carefully separated and dried.

Those he sowed directly were scattered thinly over lightly worked up dirt, patted into place, and lightly sprinkled with collected rain water until they sprouted. Others were started in plastic pots and milk jugs until time to transplant into the garden.

Porch Plants

One of the defining elements of his garden was the countless potted plants he stacked everywhere, including tender tropical plants and a few others that were easier to keep confined or to share when grown in pots.

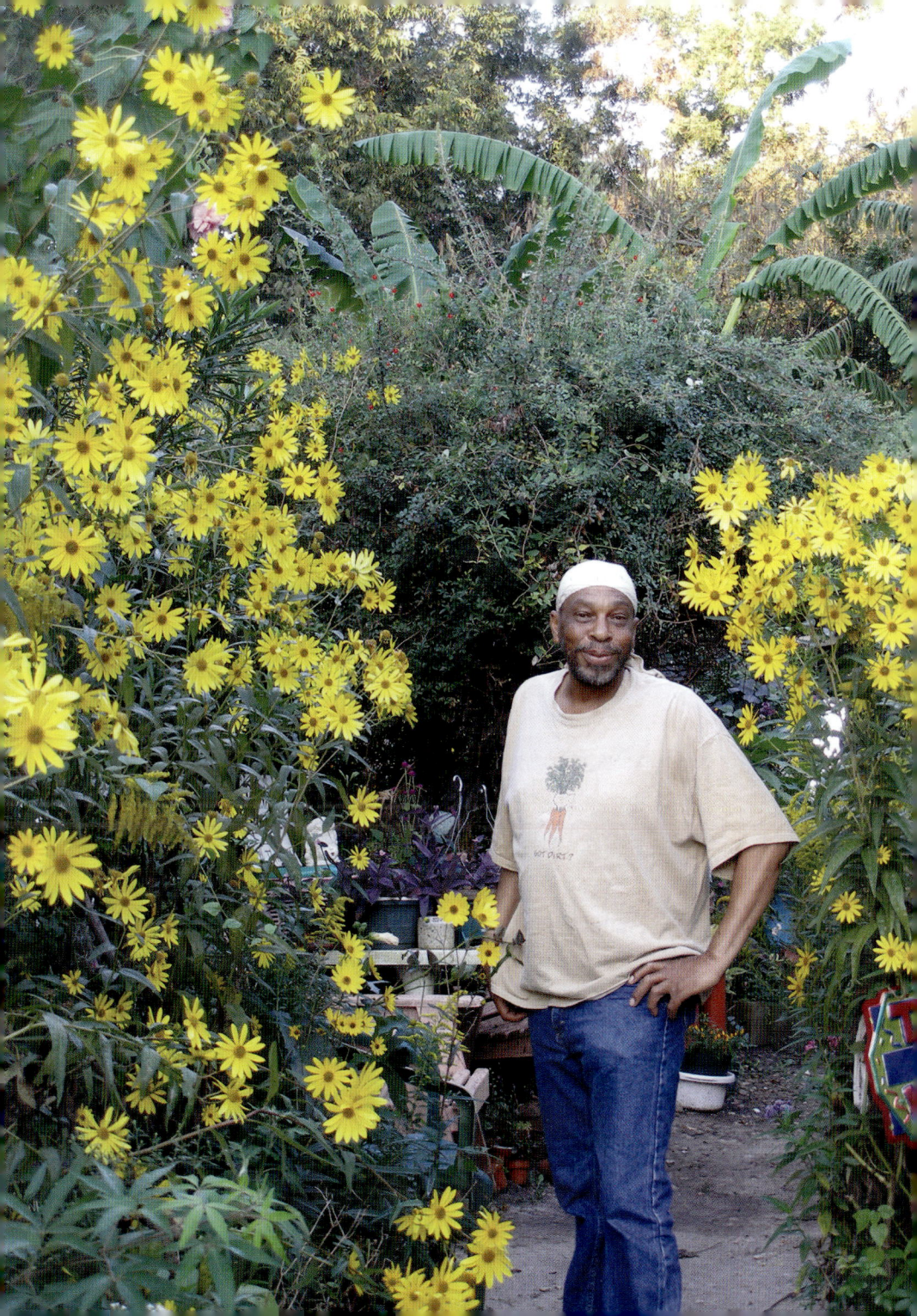

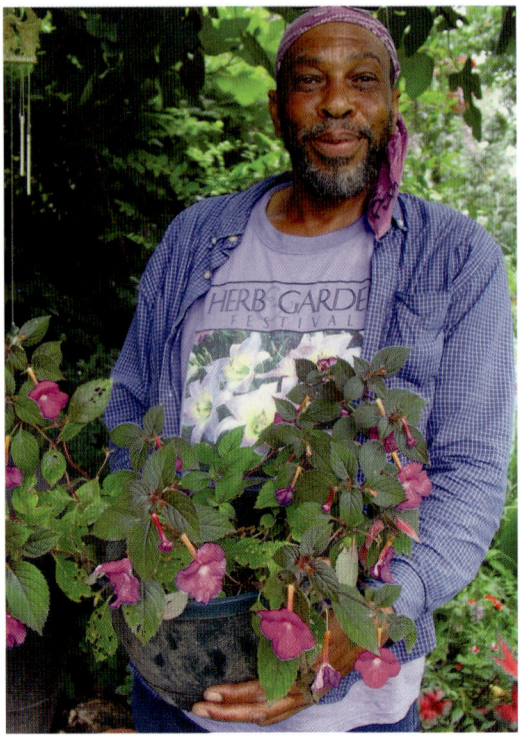

"I will plant in anything—anything. Hold out your hand and see for yourself. A flower pot is just something to hold dirt together."

Going way beyond reusable plastic pots, Dirt coaxed plants into whatever he could scrounge for free, from used tires cut and inverted as frilly containers to buckets, trash cans, mailboxes, and plastic bins. There were plants spilling from old shoes, boots, bathroom fixtures, toy trucks, athletic headgear, upended drainage pipes, and decayed hollows in his old trees.

And he often stacked potted plants atop one another so when he watered one it could drain into lower pots. When one of his ancient pecan trees was felled by a storm, its massive trunk leaning precariously against another tree, Dirt went with the flow; rather than remove it he simply piled more potted plants onto the ersatz arbor and ducked under it to get through to the rest of the garden.

Whenever someone commented that he didn't have room for any more plants, he would point out that "there's room to plant stuff right where you are standing."

He grew everything in his erratic mix of dirt and any kind of organic stuff he could get his hands on, usually leaf pile compost, cheap potting soil, or bark.

One of his prides and joys was a semi-hardy relative of African violet I identified as achimenes. Dirt and I laughed long and hard one time when a pedantic university professor chastised my vernacular way of saying that Granny used to put her achimenes "up under the porch" for the winter.

Dirt chuckled that "if Granny couldn't put things up under the porch in the fall, how could she get them *out from up under* the porch in the spring?"

Plant Thugs and Dangerous Hitchhikers

Dirt was almost arrogantly proud of his more aggressive plants. Some were overenthusiastic natives, but most were nonnatives that brought on the ire of hardcore plant proponents; some are actually banned as invasive exotics by environmental agencies.

To visitors dismayed to find him flouting invasive exotic plants, he would explain that some of his worst weeds were native plants, saying "a thug is a thug, no matter where it comes from. You just gotta deal with it."

His attitude came to a head in a classic run-in during the Q&A session after a native plant conference in Georgia, when his love of nonnative plants in his garden was challenged by another participant.

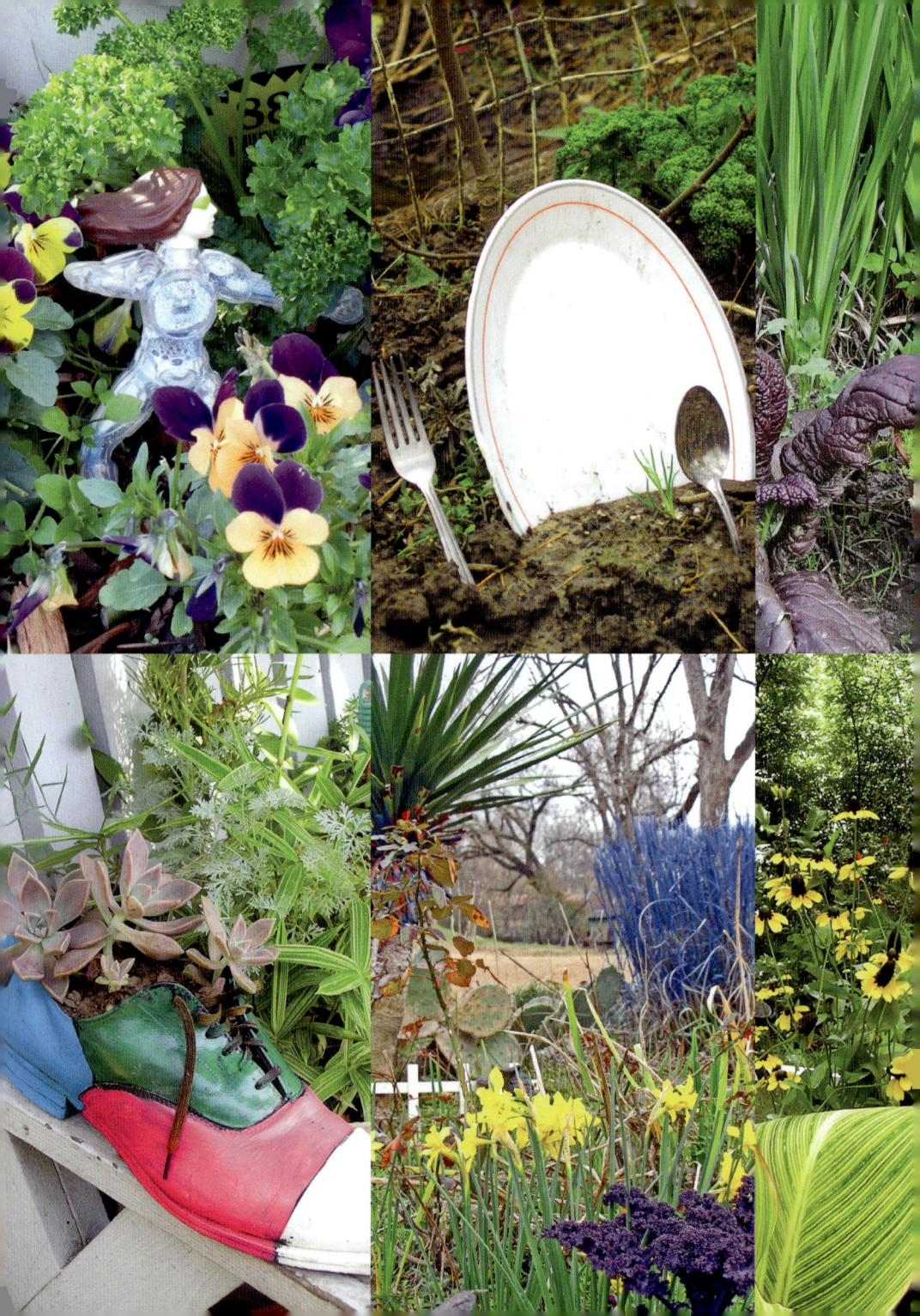

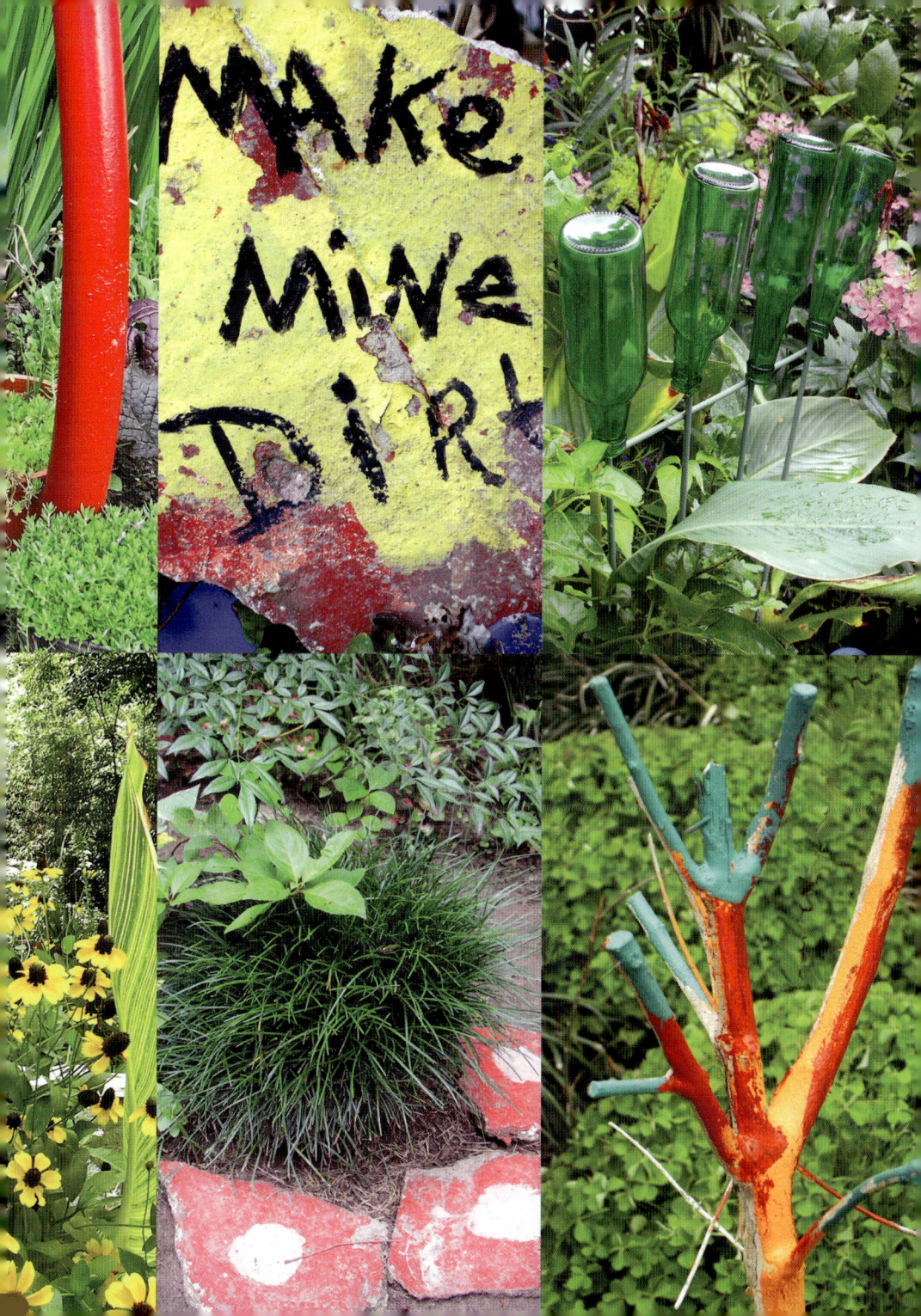

He defended his traditional use of Queen Anne's lace, English ivy, nandina, privet, and other invasive foreign species, refusing to back down from the academic accuser.

Finally, calmly standing to draw himself up to his full height, he pointed around the room, saying "You, and you, and you—all of you are old enough to understand what I'm about to say.

"We don't discriminate on the basis of country of origin. You know," he accused in his booming voice, "there was a time when people like ME weren't welcome in your community, either!"

Needless to say, the room went silent, leaving everyone to mull their own prejudices, before Dirt nailed it with "When nandina is outlawed, only outlaws will grow nandina."

In his garden the most egregious spreaders were kept wrangled through brute force, being pulled or whacked into submission (remember his Kaiser blade?). And it did no good to chastise Dirt for sharing dreaded "mimosa weed" (chamber bitter) and variegated "Limelight" artemisia along with his potted giveaway plants.

Yard Art

"Gardens are works of art," Dr. Dirt would say. "No two are alike, and they shouldn't be. Your garden should be an expression of your own self."

Characteristic of most over-the-top cottage gardeners, Dirt created and placed whimsical folk art to complement the riot of flowers. Most were found objects gleaned from roadsides or left by visitors, used singly or cobbled together into creative bricolage constructions.

The garden was gussied up with bottles, cups, dishes, broken tools, household and kitchen utensils, silvery CDs, hubcaps, and who-knows-what-else, all repurposed as plant accessories. Small dolls, miniature dinosaurs, and other children's toys peeked cheekily from unexpected perches amongst flowers and greenery.

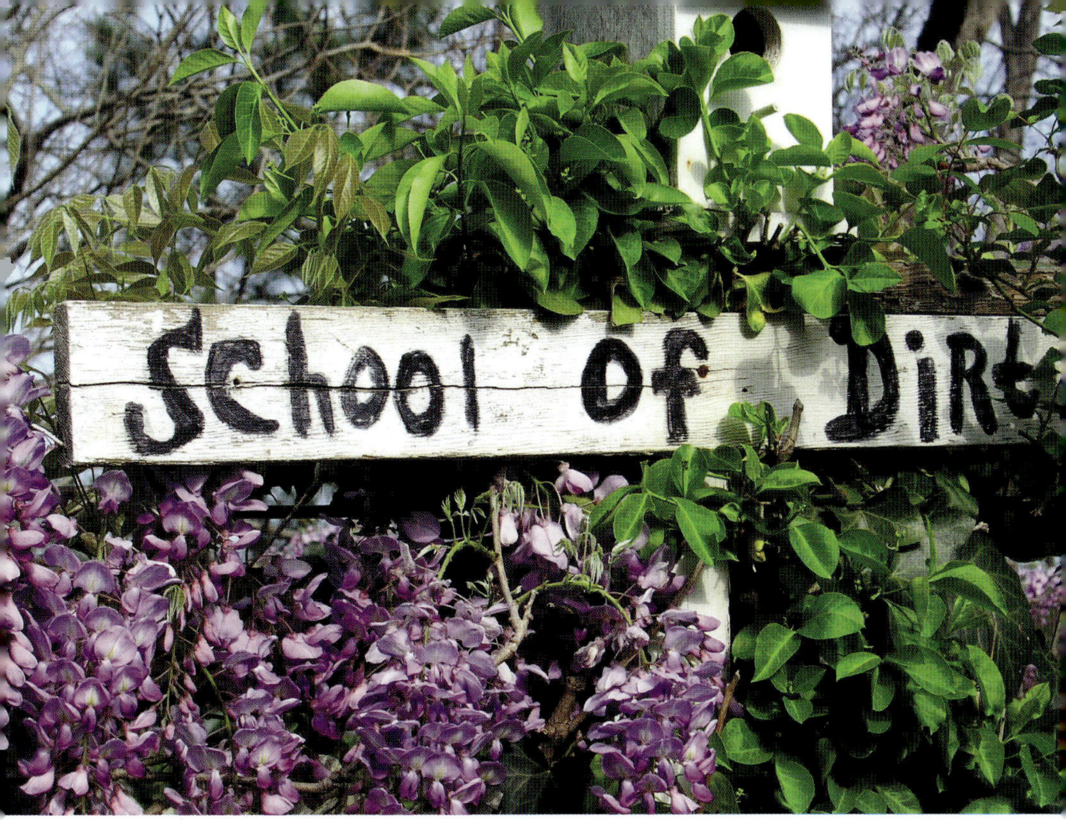

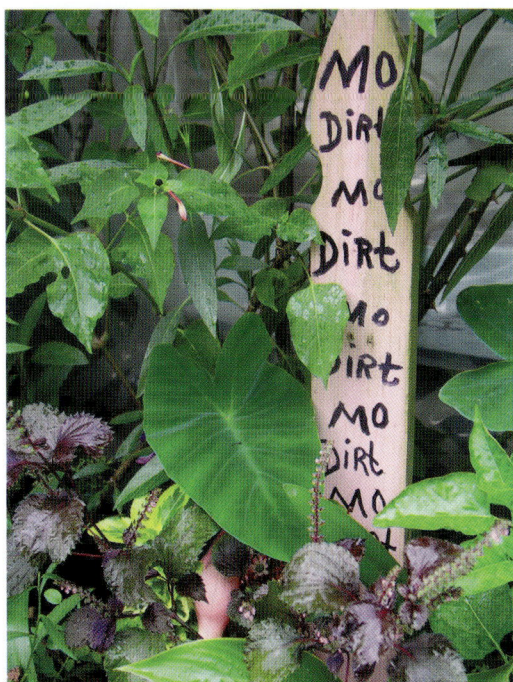

Dirt festooned the sharp tips of yucca leaves (which themselves were often colorfully spray-painted) with flower-shaped cups made from cut-up Styrofoam egg cartons. He hung crystal cups from branches, and slipped colorful glass bottles over metal rods stuck amongst his flowers.

He painted everything in vivid colors, from garden tools and old brooms to containers, rocks, sticks, the stems of dormant perennials and ornamental grasses, and even the occasional run of mismatched bricks in his walks. "It gives me something to look at, some color in the winter," he explained. Brightly lettered welcome and warning signs, and slogans such as "School of Dirt" and "Dirt Only" were everywhere.

"I've had some society ladies say to me, 'Oh, you can't mix those colors.' And I tell 'em, 'Says who?'"

Rise to Prominence

As Dirt became better known, he traveled widely across the southeast—always by bus or with friends because he never drove or flew—to appear at garden club or Master Gardener meetings, flower shows, and other garden venues, where he encouraged people to come visit his garden (and to bring plants to share).

His alternate but honest approach was a refreshing success in an era of blow-dried perfection. Newspaper reporters, magazine editors, and TV producers paid homage to both the man and his colorful garden. After filming what was planned to be a short segment within a standard program, Home and Garden Television producers turned their profile of Dirt and his garden into a full-length feature—the entire episode.

All this attention naturally brought curious individuals from all over the country, with carloads and even tour buses pulling up alongside his little corner of the world, which in turn led to his sharing—and receiving—more plants than ever.

He especially enjoyed entertaining and educating garden clubs, plant societies, and van loads of university horticulture and landscape students, who he encouraged, in his inimitably forthright manner, to embrace a more hands-on, sustainable, and historic perspective than what they may have been preached in formal classrooms studies.

"I don't like dogs or most people," Dirt unfailingly admitted to visitors, "but I love me some plants! Love the beauty of watching things grow, the smells, the whole thing. And meeting folks with new plants, old plants. Especially those some experts call 'weeds' which are just as important as any others."

"It's all about the dirt. Too many things in life aren't real." Holding up a handful of his crumbly, dark-as-chocolate soil, he said "THIS is real. This is what we come from, this is what we're going back to. Everything starts from dirt. Everything returns to dirt."

Dirt and the Professor: An Unlikely Alliance

Have you ever pondered something you experienced, but didn't fully register at the time? Looking back, you might think, "Wow, what an amazing thing happened . . ."

That's my take on our unlikely ten-year collaboration—an astonishing trip, if you will—between two men who from all appearances had practically nothing in common except being DIGrs.

I was a horticulture professor, he was a self-taught dirt gardener; we were counterpoints who, much to our surprise, added the right amount of seasoning to the other's ideas. We turned out to be kindred spirits in many ways, clown-like outsiders who got a kick from plants and knowledge.

"We share that spirit that a lot of people have," Dirt once said. "Not everybody, but a lot of people. You know the difference between fixing a fancy meal for company, or just slapping something together for family? Well, we're like family."

Budding Team

I had Dirt as a guest on my radio program several times, and in 2002, when I switched from commercial radio to the ad-free Mississippi Public Broadcasting, an affiliate network of National Public Radio, I wanted Dirt to be part of a unique new "green-neck" production.

He was an instant delight, relaxed and patient yet sparkling with comedic wit. As we blended our personal homespun garden strategies—part science, part folklore, always rife with humorous banter—"Dirt and the Professor" took off.

After a few weeks of part-time appearances, I insisted that Dirt be offered a full-time cohosting spot, and he was hired on an equal footing with me, including identical contracts.

While I guided calls and comments to stay on topic and keep our recommendations factual, he shared weekly experiences with his plants, threw in practical how-to tips, and shared simple home-cooking recipes he honed while feeding himself from his garden.

Along with our weekly shout-outs to small towns we visited, listeners were smitten with Dirt's deep, mellifluous voice and rumbling "I can dig it" chuckle.

Whenever I started getting a little pedantic over this or that, Dirt would diffuse the situation, calling me out with a laughing "Sho' you're right, Professor. Sho' you're right!"—which was my cue to calm down and relax. He often mocked me when I got mired down in book learning and got balled up in all the reasons something might not work; I admired how he talked about *just doing it*, without expecting to fail.

As Dirt would say, "It's a little of this, a little of that—the professor does the science, I do the gardening. Horticulture and gardening, like cake and ice cream."

"Felder is just a nutty gardener like I am. Yeah, he's a horticultural professor, but—he might not like me saying this—he's becoming a very good gardener."

Dirt helped me realize that it's not about rules, it's about savoring. We tried to help people to find comfort in their gardens. If you like to mow grass, or grow roses, enjoy it. We may have wagged our fingers in the faces of folks about doing things they don't enjoy, but our idea was to help people connect, and laugh, and share.

Through traveling far and wide seeking fellow gardenphiles, and sharing plants and stories with all, laughing and teasing each other with our own brand of humorous faux outrage, we transcended differences which at that time in the Deep South were fairly stark.

We occasionally heard rumors that our sense of the absurd put off a few serious-minded and formally trained gardeners, but we talked it over and decided that those were not our real audience anyway. As we intensified our teasing and cajoling to "let it all hang out," the more reticent backyard gardeners realized that we weren't critical of alternate approaches.

"That's what we try to do with *The Gestalt Gardener* radio program," Dirt summed up. "It means *the whole thing*, everything related to gardening and overlapping."

Weekly Routine

For the more than three years he cohosted *The Gestalt Gardener* program, early on Friday mornings Dirt would ride a bus from his small town of Edwards, about twenty-five miles away, to my home in

Suspicious Characters

As we prepared for presentations in far-flung areas, we would spend an hour or two driving slowly around neighborhoods getting a feel for local plants and styles.

And one of Dirt's favorite things to point out later that day to lecture attendees was how "if a scraggly, long-haired white guy in a straw hat and a tall black man wearing a du-rag can cruise your 'hood looking so suspicious, and nobody asks what we're up to, *your neighborhood watch program ain't working!*"

Jackson, bringing a fresh bouquet of flowers he had cut that morning from his garden.

Gotta give propers to Terryl, who for years put up with having Dirt and me upset her frenetic morning routines of getting kids to school before readying herself for work. Doff my hat to a real saint.

On the way to the studio, to get our garden juices flowing we would wander around my own cluttered garden to see what was going on, looking for ways to linking that to what Dirt's bouquet had to say about his own garden.

After the program, we would meander around the state looking for fellow "dirt gardeners" or making visits, then find a hole-in-the-wall café, soul food, or other locally owned eatery (where Dirt often swapped tips with the cooks), then I would drive him back to Edwards where we would wander around his garden for a bit.

That was our routine, week after week, year in and year out, no matter the weather.

Dirt's "Bokays"

To keep things personal and timely, and to provide fodder to talk about, every week Dirt brought in a bouquet from his garden—flowers, buds, evergreen foliage, dried flower or seed heads, bare branches, and the occasional artificial flower, freshly cut from his garden.

Always tied with a colorful ribbon or old Mardi Gras beads and wrapped carefully in moist paper or held in a flower vase, they were a testament to his ability to have something beautiful in his garden every single week of the year. *Every. Single. Week.*

To supplement meager cold-weather flower fare, he would include berries, dried flower heads and seed pods, evergreen foliage, interesting stems (often spray-painted for extra color), and even artificial flowers and foliage.

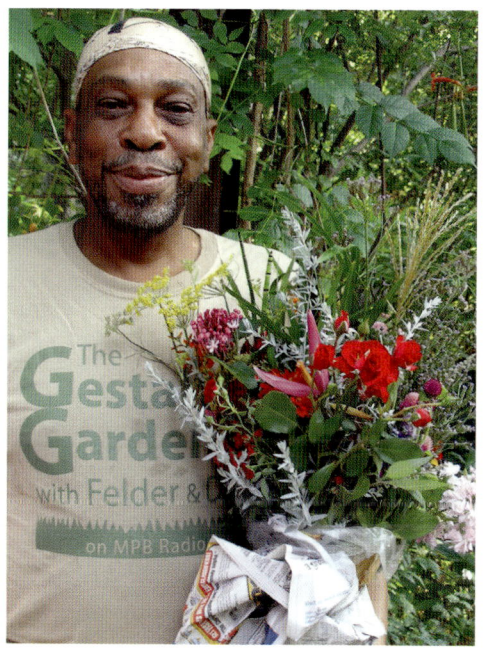
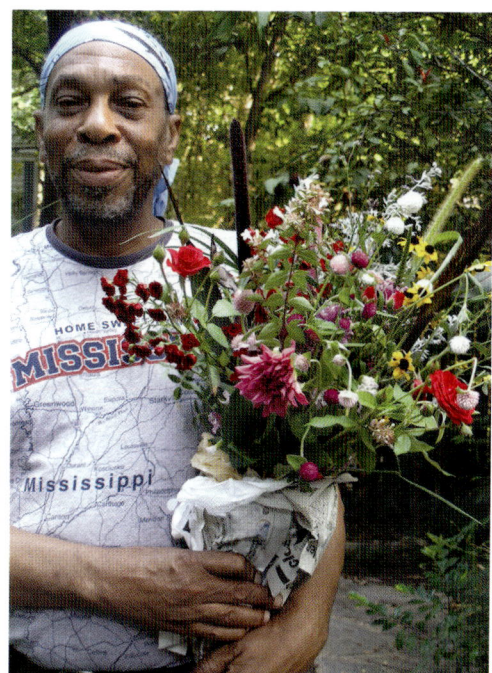
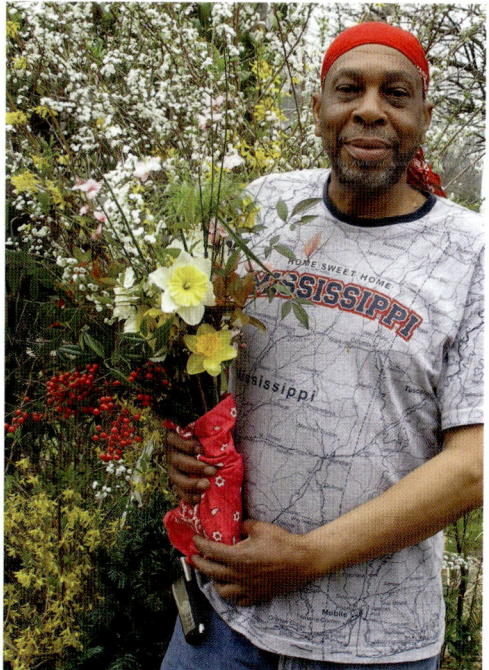

Bouquets every week of the year.

Though he had never studied floral design, he intuitively knew the concepts of line, mass, and filler; he would nearly always include "something spikey, something roundy, something frilly, and something floppy."

We would describe and discuss each item in the bouquets, which often led callers to comment on or ask questions about how they could grow the same plants in their own gardens.

In addition to being beautiful testaments to what could be grown all year, the bouquets later helped with a bit of garden forensics; poring over the many dozens of photos I had faithfully taken of his arrangements helped identify many of the plants he grew in his garden.

Farther Afield

Because we clicked so well, we ended up taking to the road, piling into my old pickup truck to appear at garden club and plant society meetings, Master Gardener conferences, flower shows, plant swaps, and general town hall meetings—anywhere "garden variety gardeners" could flock.

We tag-teamed our presentations, taking turns with the microphone while telling tales and sharing garden lore. And we usually got invited afterwards to enjoy the yards of likeminded gardeners.

Dirt and Wilma Gene

Dirt and I spent a lot of time in my own mother's garden. Wilma Gene, who spent much of her life corralling neighborhood kids to show them how to fish, catch turtles, build treehouses, and garden, often laughingly referred to Dirt as "my best son."

She once showed Dirt her ailing rosemary shrub, saying I (the, *ahem*, horticulturist) wasn't doing anything to save it. Dirt took a few cuttings, rooted them, and, in a soppily sentimental exchange just before she died, gave her one back. It was a supreme moment for us all.

He and I agreed afterwards that we thought both of our mothers were standing behind us, ready to give us smacks on the back of the head if we didn't straighten up.

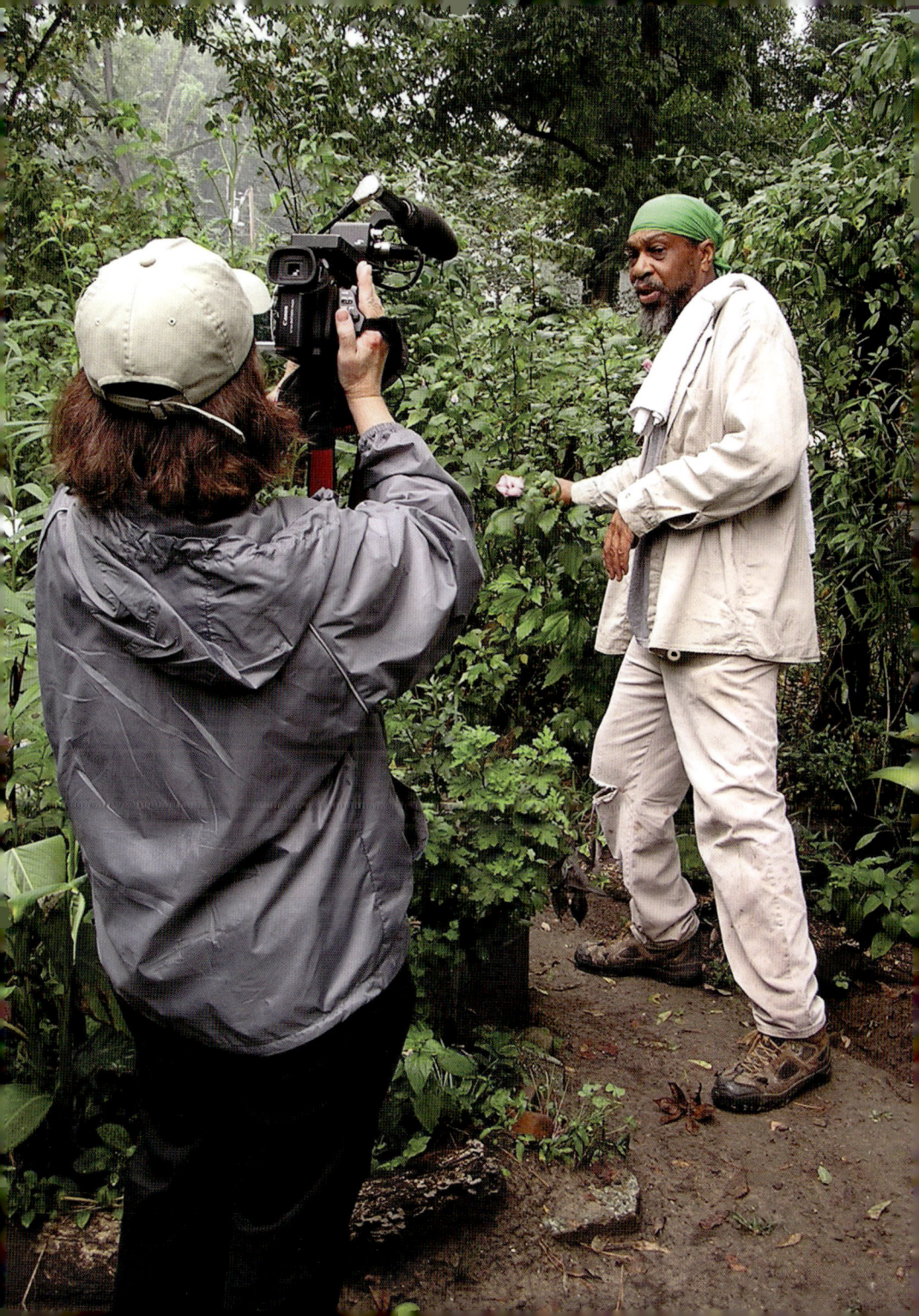

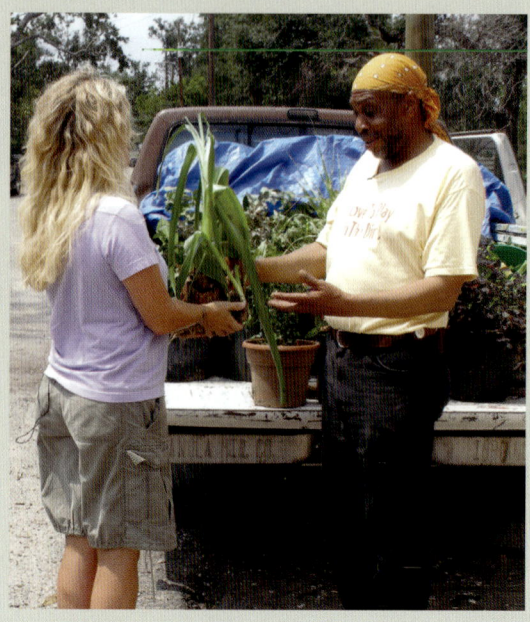

After Katrina—finding life in the rubble.

Recoastalizing

While we were cohosting *The Gestalt Gardener*, Hurricane Katrina caused deadly flooding in Louisiana, and devastated the Mississippi Gulf Coast. On the Friday before it hit, seeing what was coming, as soon as we got off the air Dirt and I jumped into the truck to tour and photograph the coast from one side to the other.

That Sunday, our own gardens, though 150 miles from the coast, suffered major damage from fallen trees and blown debris.

It took us three weeks to finally get back down, using emergency press credentials, to find . . . nothing. Mile after mile, the homes and businesses that just before the storm we had seen packed tightly on the first few blocks of streets along the coastal highway were completely gone, pushed into jumbled wreckage a mile inland.

But with the help of Steve Bender, senior garden editor from *Southern Living*, we used our radio program to organize plant donations from generous gardeners. We hauled them down, met struggling gardeners trying to reclaim what they could, and helped them start to "recoastalize" (our coined term) what had been destroyed. One of the show's listeners put us in touch with a school garden whose students were already at work replanting, months before the school could even reopen.

As we wound along our shared path, fiddling with whatever we uncovered, we met countless fellow DIGrs who also dabbled in gardening for the fun of it, and noticed how often they seemed isolated in their communities. It quickly became apparent that there was a need, in a world that typically didn't understand, to highlight their quirky or unique approaches. Dirt and I tried to help them feel more connected.

Dirt was a hit with garden journalists and producers we hung out with on our travels. His exuberant persona and folly garden provided a much-needed opportunity for otherwise-staid journalists and broadcasters to lighten up at least temporarily in the company of a colorful gardener who partied outside the box in a whimsical way and showed how folly could be a garden virtue.

He and I appeared together on some TV programs, but Dirt and his garden alone were featured in *Southern Living*, *Better Homes and Gardens*, and other garden magazines, and he was the subject of a full-length Home and Garden Television program and others.

Cultural Surprises

Because of our having marched through the Flower Power generation, and our races and appearances, when we traveled together around the conservative South and beyond we both expected the odd unkind comment. We sometimes talked about what we'd do, how we'd react if our hackles were raised, but trouble never found us.

Even while eating in rural cafes and sharing hotel rooms, we caught nary a raised eye, which we chalked up as either our team cheeriness or a testament to the slowly changing attitudes of our beloved South.

This was a good thing, because though Dirt's easy chuckles and earnest personal garden advice were lauded far and wide, much of it was a polite public façade; as anyone who spent any time around him at all can attest, the very private man wasn't shy, and never backed

Dirt's Recipes

HOMEMADE FIG PIE—Dirt's most-requested summer recipe
Find a good pecan pie recipe—one with eggs and everything.
Make it exactly like you normally would, except use pureed fresh figs instead of pecans.

HOT PEPPER SAUCE—Southern classic for using over greens and other good food
Pack small-neck bottles with clean, fresh, hot peppers.
Heat some pure white vinegar to boiling.
Pour the hot vinegar over the peppers, put the cap on, and use as needed.
Add more hot vinegar when what you have is getting used up, and keep going until the peppers get worn out.

CHA-CHA—Sauce for cooked greens or blackeyed peas
1 cup sugar
1 cup white vinegar
1 tsp. each turmeric and cumin
3 large green onions, chopped
4 large green tomatoes, chopped
Simmer them all together until veggies are tender.
Put in a jar in the refrigerator.

HERB OIL—A real treat for salad dressings or to dip bread into
Made with olive oil and fresh herbs (thyme, rosemary, oregano, basil).
Wash, then dry herbs, and stuff a few into small wine bottles.
Cover with hot olive oil, let set for two weeks before using.

PERSIMMON PIE—what else are you going to do with them?
Find a good pumpkin pie recipe, except substitute pureed persimmon for the pumpkin.
Hint: Push the fruit through a sieve to remove seeds and skin.

(Folklore Note: If you slice a persimmon seed lengthwise using pliers and a sharp knife, the embryo inside will predict the coming winter weather: a fork shape means light fluffy snow, a spoon shape means heavy snow to shovel; knife shape means cutting cold wind.)

FRIED GREEN TOMATOES—For those of you whose tomatoes never seem to turn red
Get yourself some firm green tomatoes, cut the ends off and cut the rest into thick slices.
Dredge in corn meal with some salt and pepper added.
Cook in fish frying grease, turning once (they'll float when they're ready).

TOMATO GRAVY—Basic white sauce, chopped tomatoes, sautéed veggies, seasoning
1/4 cup chopped onion
3 tablespoons bacon drippings, margarine, or olive oil
1–2 tablespoon flour
1 cup milk
salt and pepper to taste
3 tomatoes; peeled and chopped
Directions: Saute onion in bacon fat. Stir in flour until browned. Add milk, then tomatoes and juices. Stir as gravy thickens.
Add water, a little at a time and cook until gravy is the right thickness. Salt and pepper to taste. Add other herbs if you got them.

TURNIPS AND MUSTARD GREENS
Clean to get grit off the leaves, then chop and mix together.
Add half an onion and a large green bell pepper, both finely diced.
Add either a ham hock or 2–3 cups of mirepoix, if you got one or the other.
Bring it all to a boil, and simmer until tender.

down from speaking his mind. The unflinchingly outspoken man wouldn't let pass an opportunity to express his honest thoughts, often erupting with explosive rants about injustices, horticultural or social.

As Dirt summed up during a television interview, "We're going with everything we have in common, and the differences, well, we're gonna celebrate them. *Not black or white, we're just a couple of green necks!*"

Last Encounter

Dirt and I had become as close as a tall, closeted but flamboyant black activist with a distrust of people in general, and a long-haired, trombone-playing, Vietnam-era white veteran from the Deep South could be.

Both of us had spent a lot of time outside the South, seeing and experiencing different ways. And when we came home, we each realized that our neighbors, families, and friends—*our people*—needed to lighten up a little bit. And we found a good vehicle in gardening.

We had an opportunity to share and to pull people together into, if not a community, more of a tribe. And for nearly ten years we shared the good gardening life with its fun and fruits, highs and lows, and spread it as far and wide as we could.

Like most of his friends, I knew that Dirt, who usually came across as very buoyant, experienced lows that made it hard to handle the pressure of being in the public eye. Having to appear upbeat all the time, especially on live radio or on stage but also on the bus and street corner, became increasingly difficult.

But one day, like a bolt out of the blue, Dirt finally had enough.

We were both coming down after the buzz of our weekly broadcast when we were presented with our annual contracts to sign so we could continue working with NPR in the new fiscal year. It was a formality for me, but Dirt suddenly and stubbornly declined. Nothing I could say could persuade him otherwise and a hastily convened panic discussion with management fared no better.

He hung it up, retreated from the public eye to retire in his beloved garden. Our rollicking collaboration was over.

"I was happy before I ever met you, Rushing," he had frankly let me know the last time we met. "Don't like people or most animals. All I need are my plants."

And that was that. The boisterous dream team was over.

Last Words

Sadly, the seventy-year-old Leon, ailing for several years from cancer, died at home in April 2017. Even as his health declined, he continued to tend Millie's garden, welcoming visitors and sharing plants until he could no longer hold his Kaiser blade or lift a bucket to water plants.

In his prewritten obituary he assured that "As you celebrate my life and legacy, know that I am well. I was a natural-gardener known as 'Dr. Dirt' the plant doctor. I had a love for nature, especially plants.

"I had a gift for resuscitating plants of every kind. I also loved collecting antiques of all types, which many considered to be junk but were my treasure. My green thumb transformed my home place into an Eden-like garden known throughout the State and I made sure the Town of Edwards always looked good."

Eden Fades

Our greatest gardens are only a breath away from wilderness; take away the gardener, and nature takes back control. And within a few months of Dirt's passing, his celebrated garden quickly declined.

The fence was removed, exposing the bare mobile home that had replaced the quaint but crumbling old house, and the quirky but vulnerable garden art was discarded. Without their caretaker's vigilance, his exuberant plant collections devolved into a tangled hodgepodge;

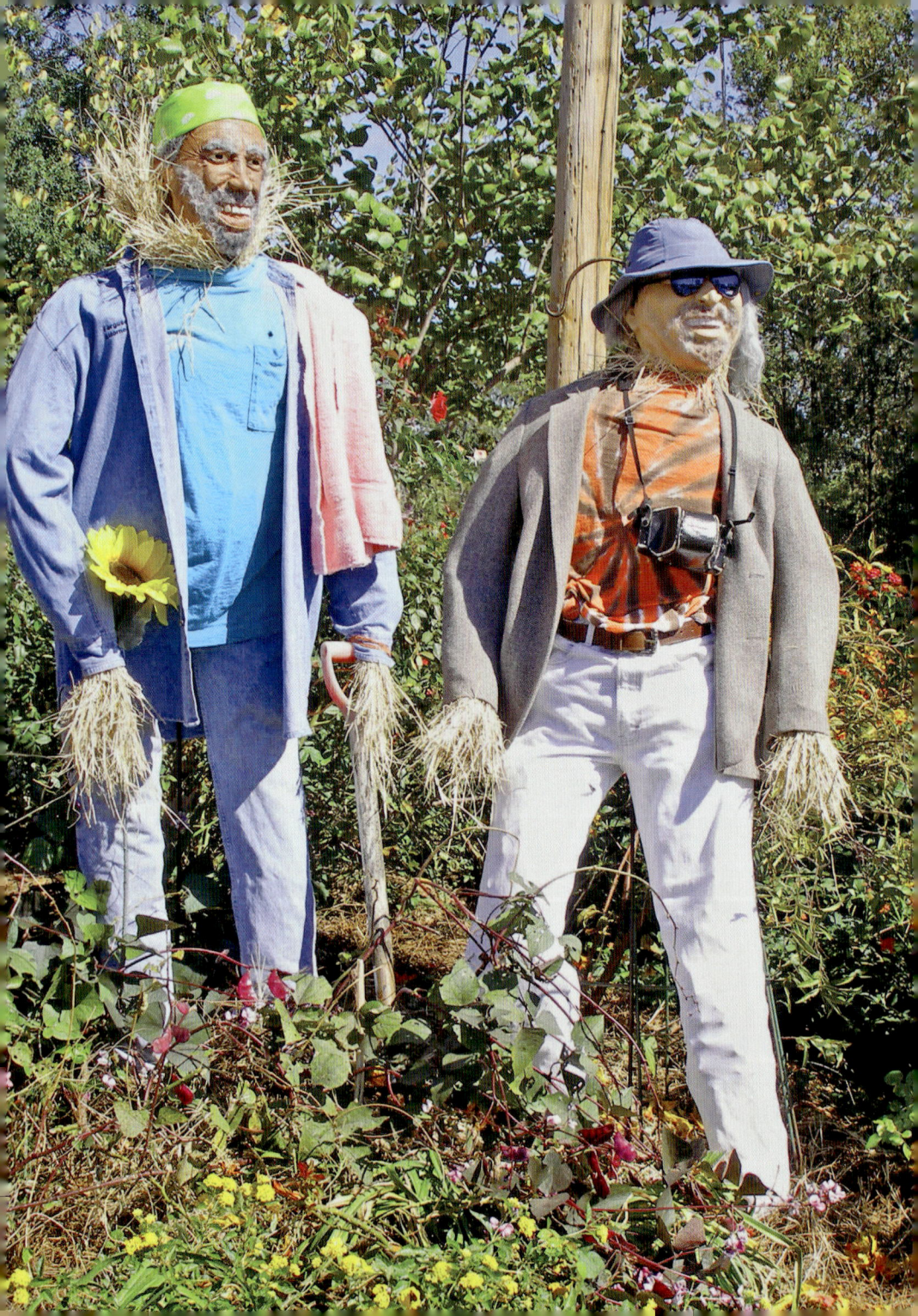

most of the garden flowers not salvaged by family and friends perished or were inexorably swallowed by more aggressive shrubs and vines that had once been held at bay with a sharp blade.

His garden is gone, and his enthusiastic mantra of "I can dig it" and deep, mellifluous chuckles are silenced. But Dirt, treasured in the hearts and minds of the every single gardener he encountered, lives on in every plant he shared.

Dirt's Plants in List Form

Note: Names are what Dirt called them; those in **boldface** were favorites which he shared a lot.

Trees

American holly
Bald cypress
Cherry laurel
Chickasaw plum
Chinese parasol (*Firmiana*)
Japanese magnolia
Mimosa
Pecan
Purple leaf plum
Red buckeye
Redbud
River birch
Vitex
Water oak

Shrubs

Abelia
Agave
Aucuba
Azalea
Beautyberry (*Callicarpa*)
Double-flowering almond
Dwarf palmetto
Euonymus
Fig
Flowering quince
Forsythia
Gardenia
Hardy lemon (*Poncirus*)
Harlequin glorybower (*Clerodendrum trichotomum*)
Hydrangea
Kerria
Mahonia
English dogwood, mock orange (*Philadelphus*)
Nandina
Oakleaf hydrangea
Pomegranate
Prickly pear cactus
Rose of Sharon
Rosemary
Sago palm
Spirea
Sweet shrub (*Calycanthus*)
Weigela
Yellow bells (*Tecoma stans*)
Yucca

Roses

Butterfly rose ('Mutabilis')
'Caldwell Pink'
'Carefree Wonder'
'Martha Gonzales'
'Paul's Scarlet' climber
'Red Cascades'
'Silver Moon' climber
'The Fairy'

Vines

Blackeyed Susan vine (*Thunbergia alata*)
Bottle gourd
Climbing roses
Coral vine (*Antigonon*)
Cypress vine
English ivy, green and variegated
Greenbriar (*Smilax*)
Hyacinth bean
Kudzu
Maypop (passion flower, *Passiflora*)
Moonflower
Morning glory
Sweet Autumn clematis
'Tater vine (*Dioscorea*)
Trumpet creeper
Trumpet honeysuckle
Wisteria
Yellow Carolina jessamine

Hardy Herbaceous Perennials

Ajuga
Artemisia, both gray and variegated
Aspidistra
Banana
Black bamboo
Blackeyed Susan (perennial *Rudbeckia fulgida*)
Butcher broom (*Ruscus*)
Butterfly ginger (*Hedychium*)
Canna
Confederate rose (*Hibiscus mutabilis*)
Daffodils (many)
Dahlia
Cut and come again (double-flowered *Heliopsis*).
Double purple **angel trumpet** (*Brugmansia*)
Ferns
Four o'clocks
Garlic
Giant striped cane (*Arundo*)
Goldenrod
Hardy gladiolus (*Gladiolus byzantinus*)
Heritage mums (*Chrysanthemum* x *rubellum*)
Hidden ginger (*Curcuma*)
Horsetail (*Equisetum*)
Hosta
Ironweed (*Vernonia*)
Lantana
Liriope
Maiden grass (*Miscanthus*)
Mexicali rose (*Clerodendrum bungei*)
Mexican petunia (*Ruellia*)

Mexican primrose (*Oenothera speciosa*)
Milk and wine lily (*Crinum*)
Mints
Monarda
Mondo grass
Montbretia (orange *Crocosmia*)
Narrow leaf sunflower
Obedience (false dragonhead, *Physostegia*)
Orange daylily
Oxalis
Painted Arum
Phlox, both woodland and **summer**
Prickly pear cactus
Purple heart (*Setcreasea pallida*)
'Purple Knight' (*Alternanthera*)
Red amaryllis
Red spider lilies, pink naked ladies (*Lycoris*)
Salvias
Sedums (many)
Soapwort (*Saponaria*)
Society garlic
Spiderwort
Stokesia
Striped ribbon grass (*Phalaris*)
Summer snowflake (*Leucojum*)
Swamp spider lily (*Hymenocallis*)
Tatarian aster (*Aster tataricus*)
Texas star hibiscus
Three kinds of **elephant ears** (including the upright *Alocasia*)
Tiger lily
Tropical milkweed (*Asclepias curassavica*)
Turk's turban (*Malvaviscus*)

Umbrella sedge (*Cyperus*)
Violets
Wild ageratum
Yarrow
Yellow flags (*Iris pseudacorus*)

Annuals

Beets
Begonias
Blackeyed Susan (annual *Rudbeckia hirta*)
Brown cotton
Burgundy mustard
Candelabra or candlestick (*Senna alata*)
Castor bean
Celosia (prince's feather, cockscomb)
Cleome
Coleus
Collards
Coreopsis (several including tall "plains tickseed" (*C. tinctoria*)
Cosmos
Garlic
Giant mullein
Gomphrena
Johnny jumpups (*Viola*)
Kale, both purple and Tuscan blue
Larkspur
Lemon grass
Lettuces
Malabar spinach
Mexican hat (*Ratibida*)
Okra

Parsley
Peppers
Perilla
Periwinkle
Portulaca, both purslane and moss rose
Queen Anne's lace
Sunflowers
Sweet potato (eating and **ornamental**)
Tomatoes, cherry
Touch-me-not (balsam, *Impatiens balsamina*)
Turnips
Vinca major
Watermelon
Zinnias

                Potted Tropical

Achimenes
Airplane plant
Aloe
Angel wing begonias
Asparagus fern
Hen and chicks (*Graptopetalum*)
Kalanchoe
Night blooming cereus (*Epiphyllum*)
Pencil cactus
Poinsettia
Sedums
Snake plant (*Sansevieria*)
Wax flower (*Hoya*)

More Keepers of the Flame: DIGr Neighbors

"People where you live," the Little Prince said, "grow five thousand roses in one garden . . . yet they don't find what they're looking for . . ."

"They don't find it," I answered.

"And yet what they're looking for could be found in a single rose, or a little water . . ."

"Of course," I answered.

And the little prince added, "But eyes are blind. You have to look with the heart."

—**Antoine de Saint-Exupéry,** *The Little Prince*

Independent gardeners are widely scattered but can be found in every community, across all social lines, usually connected through the backchannels of shared plants.

Their approach isn't always about solving problems; they enjoy the garden as a day-to-day series of experiences. Theirs is the journey more than a destination, and they keep it simple, just like the two tips I love sharing about growing plants: Dig a hole and put a plant in it, green side up. That's it." And "Some plants don't work right, or die. Replace them—that's called gardening."

Rather than drop names and places of the countless DIGrs I have met around the world, I'm going to represent most of them by highlighting several who live within walking distance of my little

Mississippi cottage garden and whose inner sanctums encapsulate the DIGr approach.

They include a renowned landscape architect with a quirky personal garden, a retired nurse who tends a Caribbean cottage garden, a collector who works at a garden center, a grief gardener, and an urban guerrilla gardener. All of them hold up their heads as they follow their own garden instincts as opposed to what others expect of them.

There are others, all within walking distance; again, they could be from anywhere; I'm sure you know at least one close by, no matter where you live.

Rick Griffin, Pied Piper of Personal Expression

Favorite sayings: "Oh, WOW!" and "No, it's *perfect!*"

I've always said that my longtime friend Rick Griffin's fine-tuning is off a little bit—but in a splendid way.

The celebrated landscape architect sees things not just from outside but from above the box; from this perspective he conjures useful solutions to create highly personalized spaces with innovative twists.

And when prompted, his ideas spill out like a popcorn machine in overdrive.

While the professional Rick Griffin works with whatever style his client prefers, it's a whole 'nother story when he gets home and allows his self-confessed "natural inclination to the funky" full rein.

It is simply part of his subconscious. "Like picking out art, I can't really explain it the same way twice. But it's like shaving yourself, or letting someone else shave you, or you shaving someone else. You do your own face easily, but just try shaving someone else, or letting them shave you."

In Rick's idiosyncratic personal garden, he practices what he preaches, gleefully trying out his wildest, more in-your-face concepts

Keep Fondren Funky

Soon after the Civil War, a small town was carved from destroyed plantation farmland by Isham Cade, an ex-slave who subdivided the land into parcels. Located next to the original 1850s Mississippi Lunatic Asylum, for years the Isham Cade Survey was known informally as 'Sylum Heights.

Now called Fondren, the village of about 2,500 peaceful residents is celebrated statewide as eclectic, welcoming, and uniquely diverse (racial, social, educational, and cultural). Its gardens are a rich tapestry of styles woven with traditional Old South, art deco, 1940s suburban bungalow, naturalistic woodland, picket-fenced cottage, and front-yard vegetables. Boasting the world's heaviest concentration of glass bottle trees, its motto is *Keep Fondren Funky*.

It's hard to tell which came first—did DIGrs inject funkiness into Fondren, or did Fondren's quirkiness lure them in and make them feel more relaxed?

Whatever. Liberated Fondren gardeners feel welcome enough to express themselves through yard shows. Here are a mere handful—there are others, including an architect and glass artist who grows vegetables in his front yard; the guy around the corner with an ever-changing carved wood art display, and the couple whose wonderfully diverse naturalistic garden irritates their next-door neighbor's more structured sense of propriety.

Bottom line is many of my nearby fellow gardeners do their own things, in their own styles, regardless of what neighbors may think.

It's hard to feel isolated, no matter how off-kilter, when they can watch other DIGrs' yards being accessorized just around the block, and find camaraderie by sharing plants and stories out in the street.

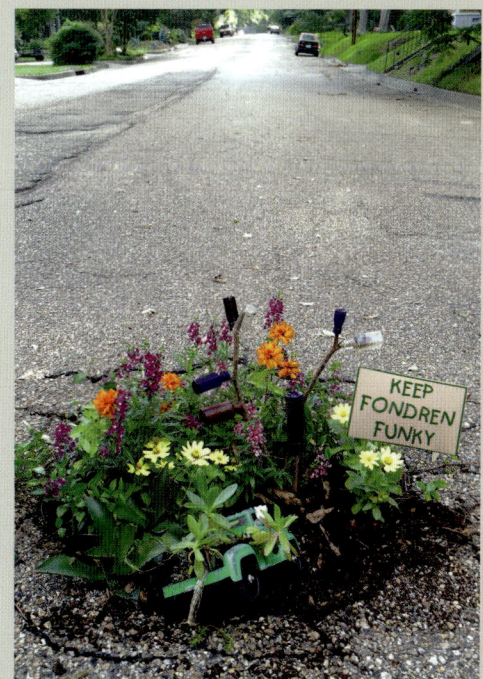

Taking "Keep Fondren Funky" to the streets.

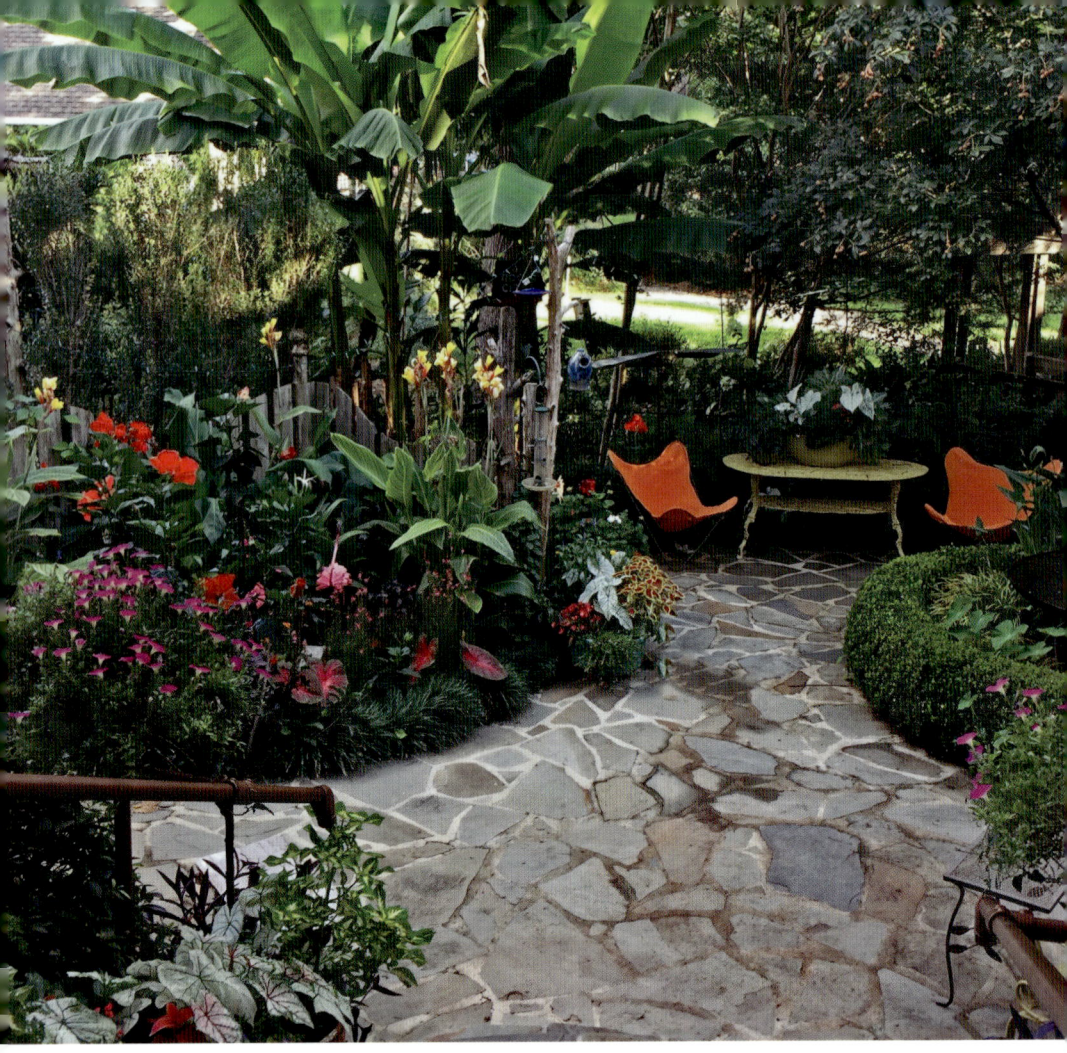

and showcasing to students in his college landscape classes that sound design principles can work even when carved from nontraditional materials.

Rick says what he calls his garden carnival is designed first and foremost for living, relaxing with his family and hosting guests. But it's also his playground where art and even butterflies, birds, and dragonflies are welcome to help create theater.

And he delights in getting away with using whatever catches his attention, though admitting to treading lightly in his and his wife Shirley's upscale but conservative neighborhood whose homeowners are baffled by public displays of eccentricities.

Always with an eye for creating useful but comfortable, highly personal outdoor rooms, front and back, he embraces smaller, well-shaped lawns and almost obsessively weaves a rich tapestry of contemporary—some might say quirky—elements on sound foundations, contrasting shapes, textures, and colors in materials (including recycled) and tried-and-true plants.

Rick and Shirley bring the same qualities indoors, with hand-painted floor coverings, corrugated tin ceilings, paintings and broken-mirror mosaics on walls, and collectibles and one-off *tchotchkes* everywhere. Even the dining room chandelier is crowned at Christmas with a twisted, thorny branch festooned with gumdrops and beads.

The relatively small garden has an ingenious meandering feel with winding paths and lavish plantings and arbors that conceal hidden areas. Guests coming to Rick's patio porch are treated to an eclectic mix of plants all year round, including tropical delicacies, roses, herbs, annuals, and perennials. Wandering through the various rooms of his garden they come across seating, sculptures, wind chimes, a bird bath, bird feeders, a metal screen, and a small greenhouse that doubles as a chapel – stained glass and all.

His flagstone walks are of salvaged broken sidewalk concrete stained with subtle glowing colors and embedded with round stones, marbles, and occasional flush, neck-down wine bottles; his fences, usually of well-seasoned wood, rusted woven wire, and artistically bent welded reinforcing rods (sometimes driven over by his truck for more pronounced curves) are salvaged from old homesites.

A sturdy retaining wall was created from stacked car tires filled with dirt, camouflaged with paint in natural colors, and knitted into

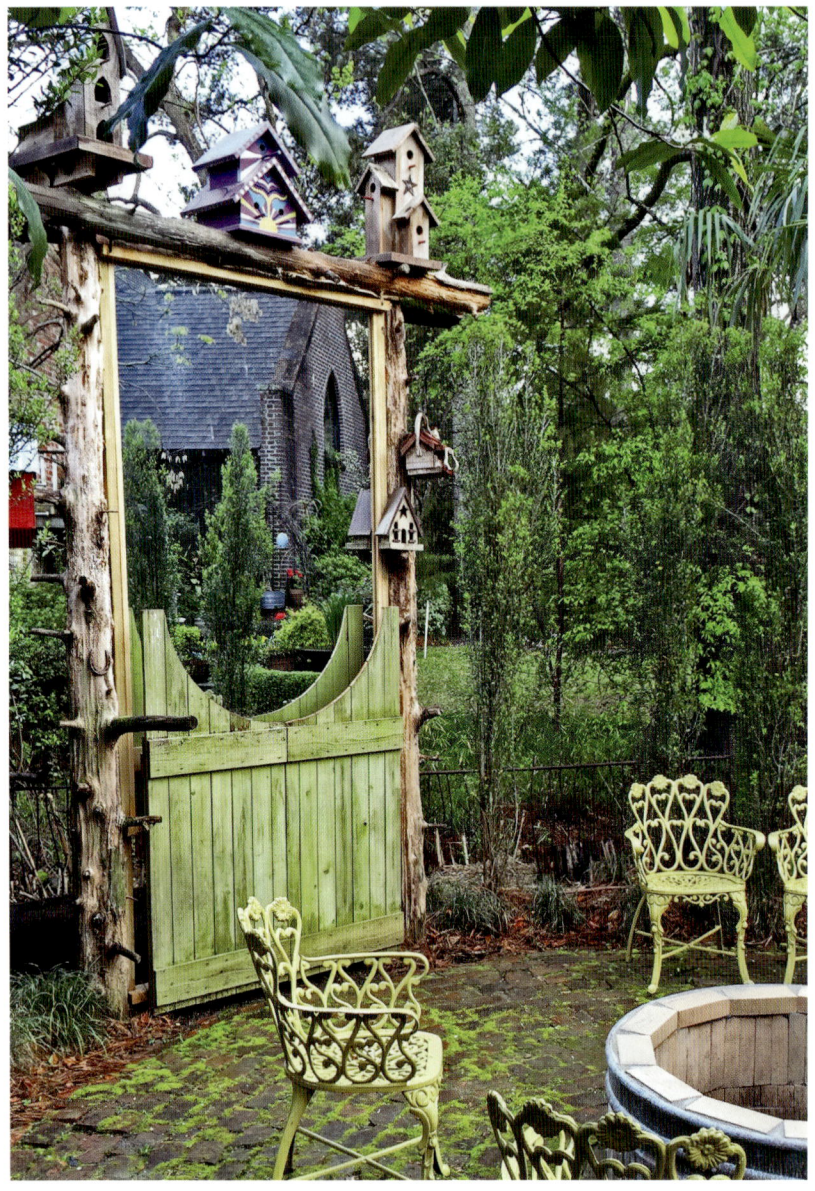

the hillside with spreading shrubs and groundcovers. To protect their small courtyard and raise its winter temperatures a few degrees, Rick and Shirley spent an entire weekend building an English-style stone wall, using stones of different sizes to give the wall a relaxed, organic feel. As Rick says, "Luckily, it's not supposed to look perfect."

Taking advantage of how imagination can be part of a landscaper's tool kit, visual hints lead the mind in unexpected directions. Rick, ever the landscape omnivore, takes advantage of and enhances "borrowed views" to create the feel that his garden is bigger than it is. Paths, arbors, bridges, and gates added around the fence give the illusion that they lead into other garden rooms, when in fact they don't.

He built a vine-covered arbor and gate towards one neighbor's garden and a short slate path towards another as visual devices. "If only one of my neighbors had a wildflower meadow to attract butterflies and bees." Rick says. "I'd love to borrow that kind of view and any passing pollinators would be welcome too!"

And when his daughter Amy said he needed an entrance to the east for good feng shui, Rick fashioned a totally realistic faux gate by framing an oversized mirror between foraged rugged cedar trunks, adding boards at the bottom, and topping it all with hand-made birdhouses. Reflections of the garden make it appear to be a window to another garden.

He once collaborated with me on a challenge I threw out, of designing side-by-side micro-gardens with all the trimmings—covered patio with seating, open "lawn" area, a walk, plants, water, and art—at a downtown festival. The caveat? *They had to fit inside a single parking space.* His appeared much larger when he directed his path to connect with mine.

And to put it mildly, Rick is exuberant when it comes to container plantings. Pots of all sizes and materials, *everywhere*, are enablers for his uncontrollable cravings for plants, allowing him to pack in more all-year color. In addition to stacked collections of assorted clay pots (often

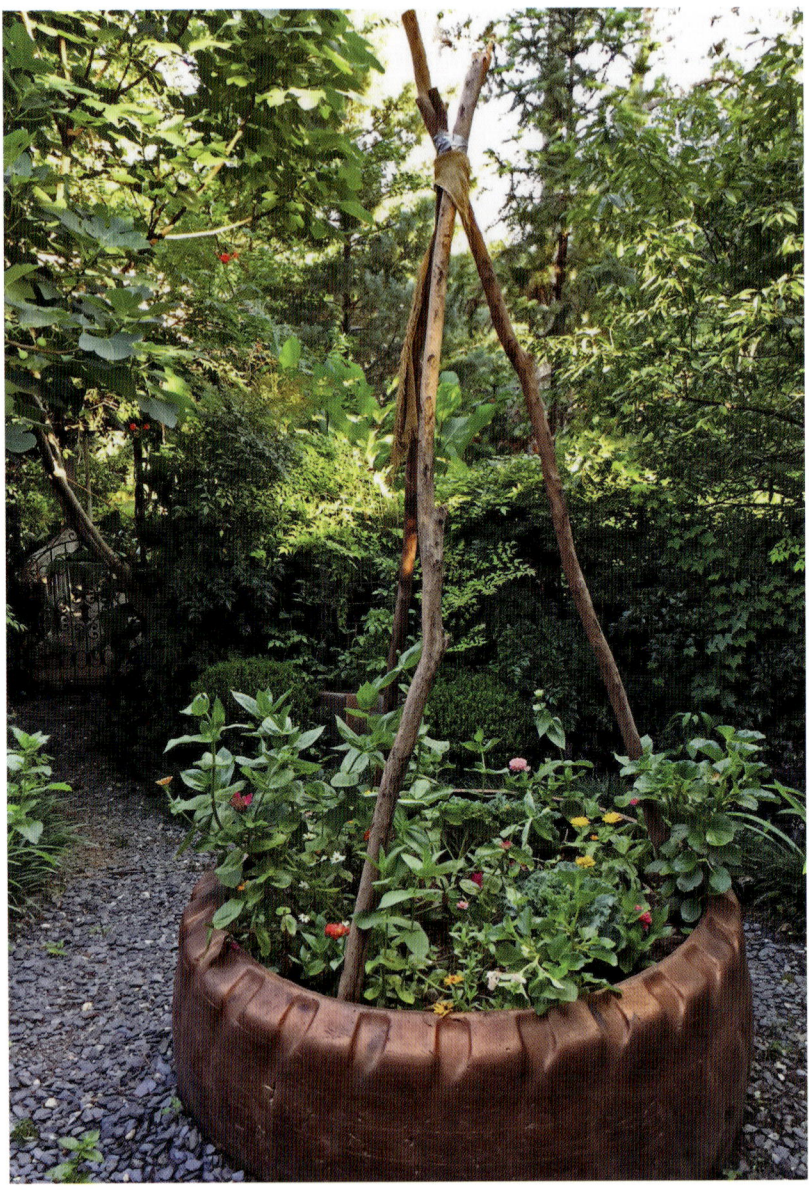

hand-painted abstractly) he has a large metal horse trough, perforated washing machine drum, inverted used car tires, and a large, rib-tread tractor-sized tire painted copper planted with vegetables and herbs.

Vegetables, herbs, and cut flowers, all which Shirley uses indoors, are tucked everywhere. And though he regularly drags cast-off plants gleaned from remodeled landscapes, and wildflowers from fencerows, he eagerly tries new plants he comes across at horticultural trade shows or in regional magazines in his garden; those that do well he works into client designs.

Art

Rick insists that everybody has a little bit of artist in their inner child, from the most well-mannered, laid back, polite people to the type who, as he put it, "bounce while they talk." And whether they expose this through art that is homemade or artisan, it should be a deeply personal display.

"I'm *always* looking for stuff for my yard, but I don't go out actually looking for specific things. In fact, when a new space begs for an accent, I leave it vacant, and wait until the right piece presents itself to better express how I feel or think at the time.

"I used to think art had no function, not like a sprinkler system or arbor. But it can also add texture and contrasts, like using a solid scarf or belt with a patterned dress. It's shiny color in a shade garden. That's the secret in design—contrasts."

Rick shared a humorous insight, saying "I'm a pattern freak, my personality loves busy, busy lines; that's my individual taste. But I also love heavy, solid things, just not slick. Take my gate for example. It says everything about Rick Griffin. I made it clunky and drab, and durable, and I like how I designed it and how it works.

"But for three years, for some reason, it seemed just awful. Then one Sunday morning it suddenly came to me that I could paint

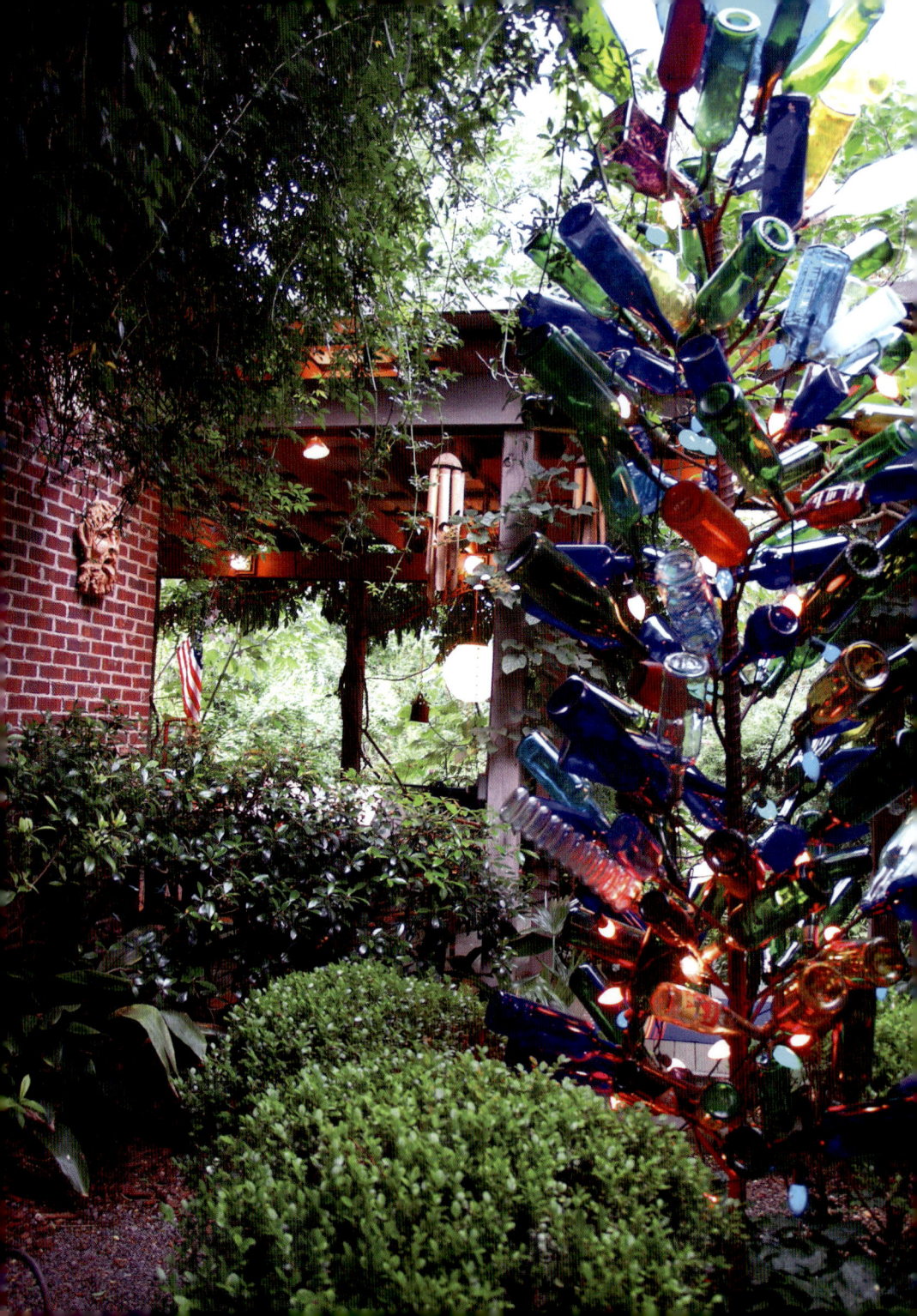

freeform colors on it! And I had to do it, couldn't leave it alone. Shirley was getting ready for church, and—God, she hates it when I do this—I was out with the sunrise, still in just my underwear, painting the fence . . ."

Becky Potts, the Gatherer

Metal, glass, pottery, fabric, and mixed media art lend an extra special dimension to Becky and Don Potts's home, indoors and out. Much of the funky art hails from their many artistic friends, turning the Potts' garden into an outdoor gallery.

Becky gave up a college teaching career to follow her husband Don's work across the country, and she found herself working part time with interior plant maintenance firms. "When we made Fondren our home back in the late '80s, Don started planting vegetables in organic raised beds. I tucked a few flowers in here and there, and started collecting succulents and potted plants."

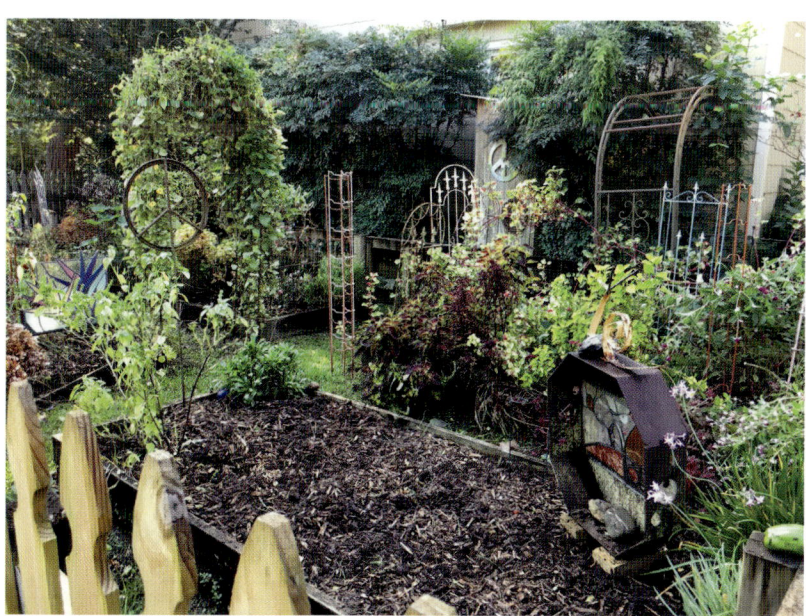

Rick Griffin Garden Design Tips

- Make your garden an all-year people place, with outdoor seating, protection from sun and rain, lighting, even fans to keep air circulating in hot or mosquito weather.
- Use circles and curved lines: "Nothing in my garden is straight. God made things round: man made them square."
- Make the garden busy with plants, but leave open places where the eye can rest.
- Every garden should have three ingredients: fire, water, and art.
- Decorate garden walls just like you do in your home. Color everywhere!
- Always be on the lookout for architectural stuff you can use or adapt. You can incorporate a lot of different objects and material, if you soften them with plants.
- Wood, stone, rusty metal, and other naturalistic materials can be more decorative than mass-produced, uniform materials.
- Keep experimenting—a good garden is never finished, always changing. Switch things up to experience the garden more vividly.
- Don't set out to control the garden totally. Relax, see what works for you and what needs help. Keep nudging it in the right directions.

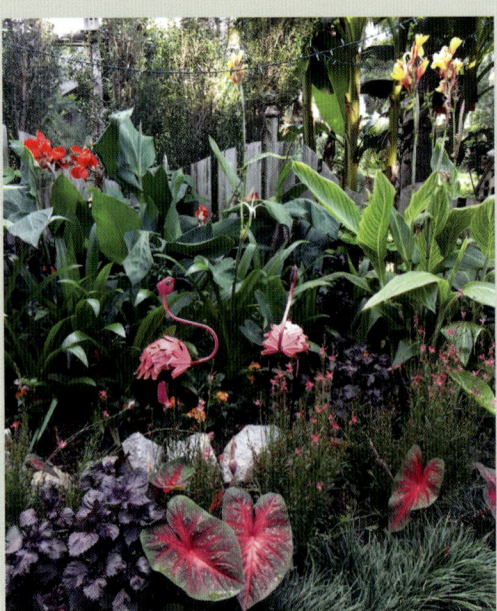

As Rick put it, "If you're not enjoying the time you spend in your garden, something's wrong."

Don owned up to being a simple foodie, but said it didn't take long for flowers to take over and spread to all corners of the garden. "Still," Becky said, "I guess that nine out of ten visitors ask what's edible and pretty much ignore the rest."

Becky bugged the greenhouse manager at a local garden center about gardening tips so much that he finally told her if she wanted to know everything he did, she'd have to start working there. "It opened up a whole new world and gave me first pick of all the cool stuff that comes in. It supports my plant habit and is my best opportunity to teach people about plants—that's my real passion."

Becky is a self-described obsessive plant collector. When asked how she decides if a new plant is worth stuffing into her already overstuffed garden, she just laughed. "When I come across a plant that's variegated or miniature, *it's mine*." Her well-used potting shed pays tribute to how much pleasure she gets from the quiet process of

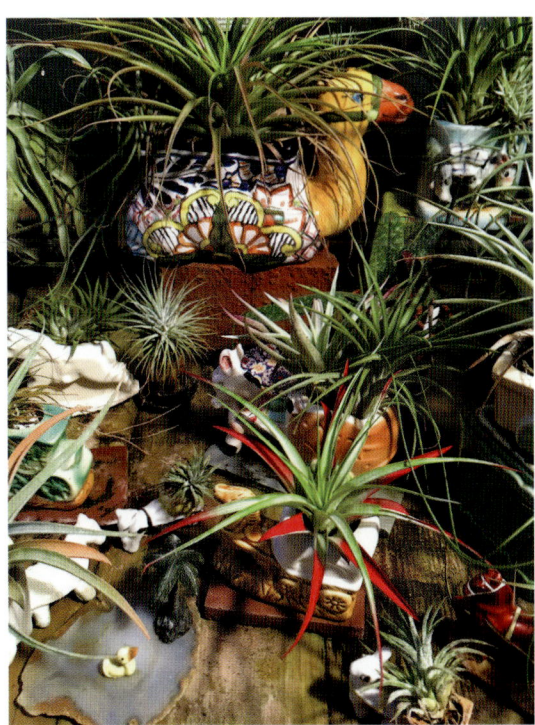

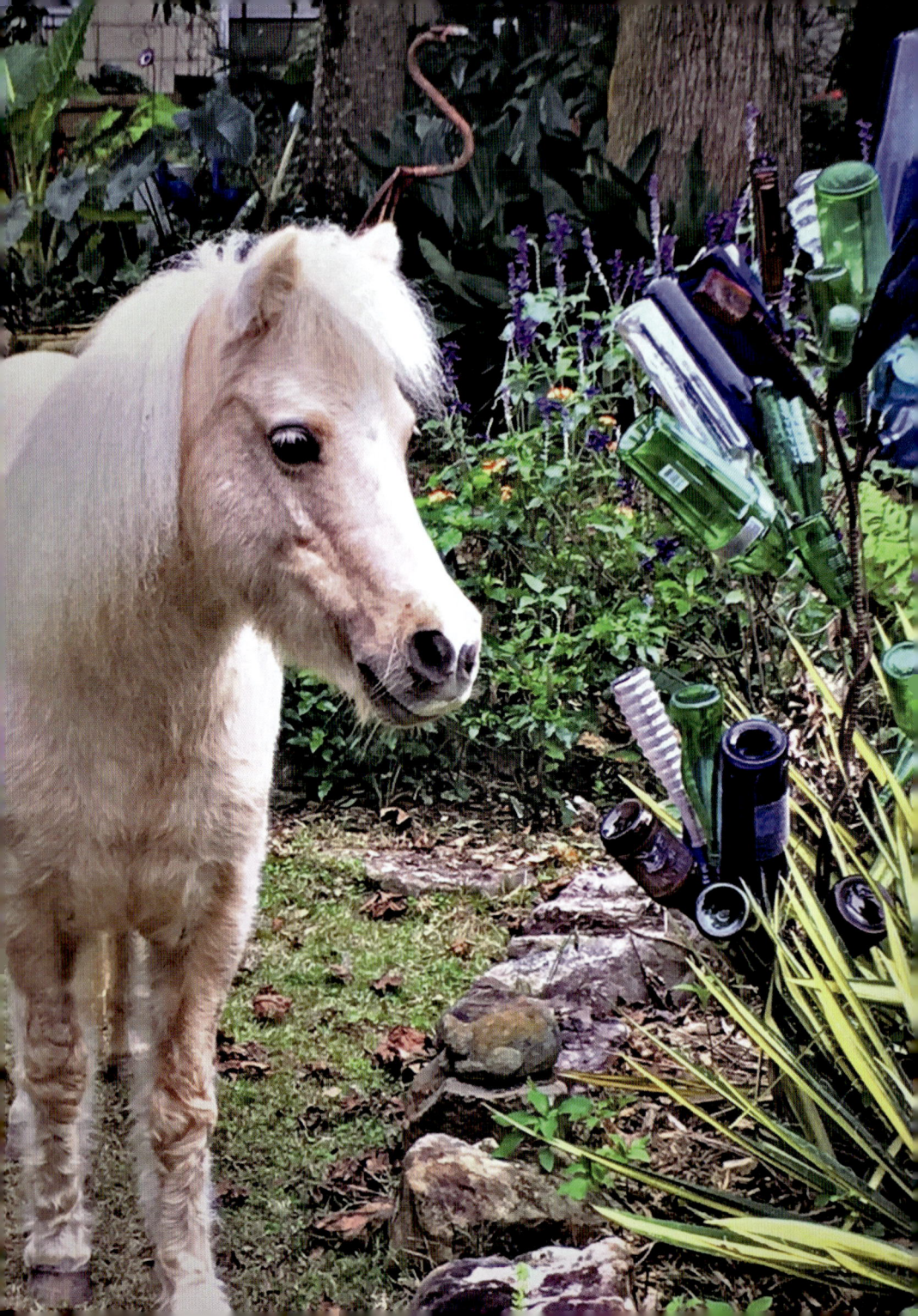

propagating plants from cuttings and seeds. "My small greenhouse barely fits all the ones that won't make it over the winter, so I root a lot of cuttings to save the plants. I love big, bold textures and colors, which explains all the coleus and elephant ears."

For many years now the garden has been a magnet for both children and grownups who want to pet their real-life miniature horse, who has free rein to reign as queen of the garden. "First thing I do when I get a new plant is show it to the horse. If she takes a bite, I plant it outside the fence."

Being able to tread lightly means a lot to the couple: their garden is run on organic principles. And both the garden and small greenhouses beg closer attention: at first glance vignettes of plants and accessories all but run together, making it hard to see the trees for the forest. A visitor is forced to slow down and linger to let each scene come in to focus—the garden's standalone "floating beds" are higgledy-piggledy plant combinations given focus by art; greenhouse benches are packed with dozens of side-by-side scenes of tiny plants in quirky containers, each group accessorized with miniature gazing balls and figurines; over all, the benevolent presence of the horse motif.

When asked what motivates her, Becky looked back in wonderment at the question, then simply said "You just gotta garden. Don't you?"

Which makes her a pure-hearted DIGr.

Suzie Cranston, the Birdhouse Lady

There are birdhouses in my neighborhood, but only one *birdhouse lady*.

Suzie Cranston, overcome with grief over losing her young son, Peck, found solace in an inspiring garden sign saying "Peace begins in the garden" and created a special tribute to his memory. Within and without her neat picket-fenced garden are everblooming roses, daylilies, purple coneflowers, gaura, lantana, camellias, towering sunflowers, ferns, and more. Much more.

"I have plants from Dr. Dirt's garden," she says, "from a friend of mine who lived near him and knew him really well. There's something in flower every month, and the bees and butterflies and hummingbirds seem to never leave. I had a sweet neighbor tell me one time, 'Dear, you're going to cause a car wreck on the street with all those beautiful flowers.' But I keep at it, planting and watering and weeding. That's what I love about gardening—you can't be bored or feel sorry for yourself because there's always something to do."

The most striking thing about Suzie's garden, however, are the birdhouses. She started adding unique collectable birdhouses found at yard sales, flea markets, and from local craftsmen. There are easily a hundred now nestled among the year-round flowers; some are painted gaily with bright and pastel colors and polka dots, while naturalistic ones, made out of tree roots, are suspended from an oak tree, turning slowly in every breeze.

Surprisingly, given a plethora of tasty choices, there are few nesting birds to be found. "An artist once did a watercolor of my birdhouses, and wrote in the notes that he wondered how a woman with this many birdhouses also loves cats. Maybe my cats are why I don't have birds in the houses . . ."

Suzie, eager for others to enjoy her happy space as much as she does, welcomes visitors with a broad smile and points out things they may have missed: unusual flowers, garden art, a new birdhouse, and the things that Peck would have loved, such as the tortoises which appear everywhere. And her sincerity shines as she says, "My yard makes me so happy . . . and people stopping by to visit it makes me even happier."

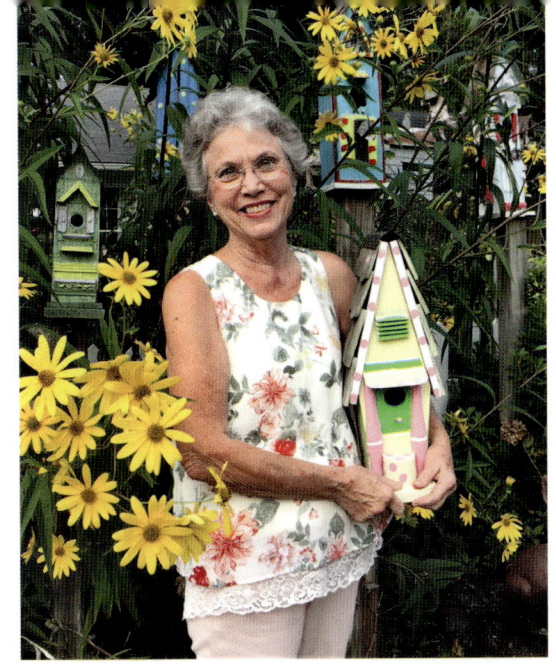

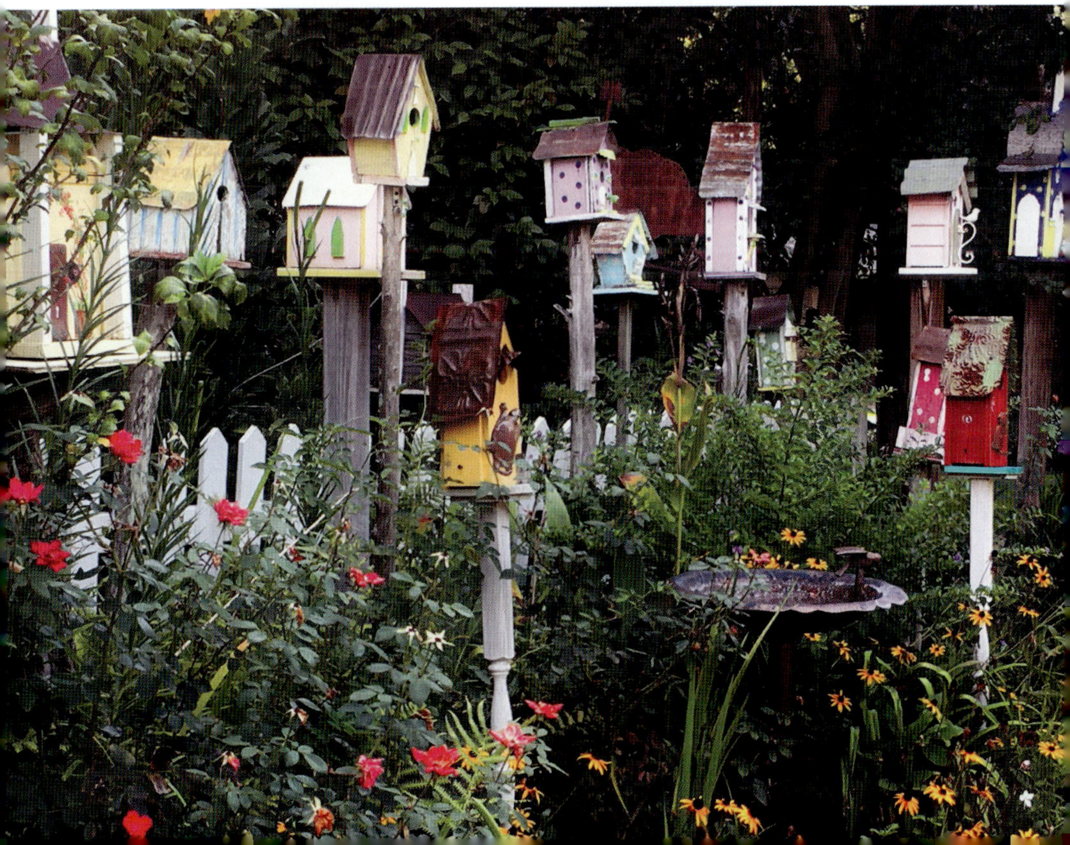

Jenny Nelson, Sharing Spirit

Every time I approach my bathroom mirror, I see a faded but uplifting note tucked to one side. I found it several years ago, slipped under the windshield wiper of my truck that has a quirky garden planted in the back.

The short message scrawled on the back of a grocery store receipt was simple but stirring: "Truck is lots of fun. So nice to be different. Jenny" Took me awhile to figure out that it had to be from a neighbor from several streets over named Jenny Nelson, a retired nurse in the mental health field who emigrated to the US from her native Jamaica.

I had met her by chance, after doing one of those immediate "wow—turn around and go back" garden discoveries. My young son Ira was with me at the time, and I like to think visiting gardens like hers—and meeting gardeners like her—helped form his worldview.

Jenny wasn't surprised at all to have impromptu visitors. Said it happened all the time, as she welcomed—with a joyful hug—our admirations of her flowers.

Unlike the tightly clipped, house-hugging landscapes up and down her street, Jenny's was an avalanche of shapes, sizes, colors, and textures, all the way to the street. Towering perennials leaned over blousy shrub roses, hydrangeas, weigelas, and other old-fashioned shrubs, with hardy flowering perennials including bananas and other tropicals and exotic annuals poking through it all and cascading over the curb.

When I said her garden "is our neighborhood's most colorful all-year Christmas tree ornament" she laughed, and with a graceful wave of her hand said in her melodic Caribbean lilt, "Everything grows in my yard—I'm sure my neighbors would be pleased to see some of it go!"

She went on to explain that her exuberance is a family roots thing, just part of who she is. Her creative spirit is quickly jaded by flowers planted in uninteresting neat rows and boxes, so she is more comfortable in what she calls "my Mississippi jungle" and its more lush,

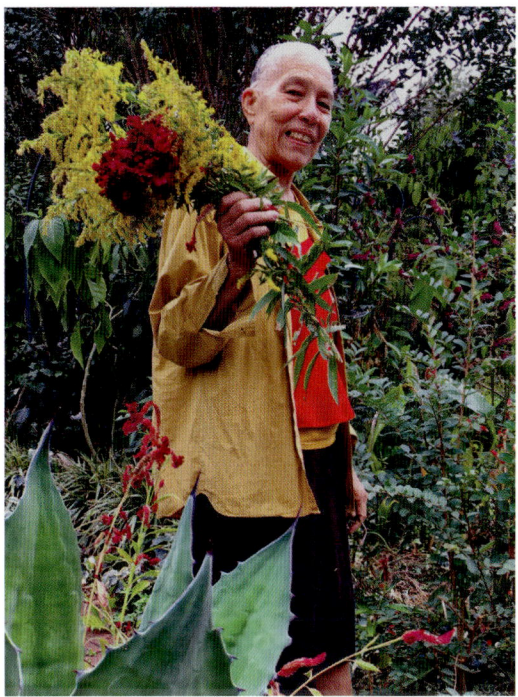

enveloping style with lots of big leaves and flowers, and unusual potted specimens.

"I have lost some of my tropical plants over the years to sudden freezes, so when it's going to be cold, I have taken on the work of bringing some that are dear to my heart indoors no matter the size of the pot. And the occasional lizard and tiny frogs do hitchhike indoors—but creatures come with the territory, don't they?"

Typical of many cottage-style gardens, hers faces the house, not the street. "All sorts of things go into gardening that console us—vitalize us, help us find whatever we are looking for from life. So my garden is planted not like the others, but where I can see all of it from my house.

"When the weather is nice, I can sit out in my jungle as long as I want and in any kind of attire, and everybody walks by but nobody sees me even though I am very close to the street. And of course it's the flowers I want to see anyway, not the people passing by!" she laughed.

On rainy or what she called "emotionally not-so-healthy days" Jenny sits in her window and see her plants and rejuvenates herself. "Though walls separate us I am still with my plants. It's also the days when I make decisions about what goes and dream about new things to come."

Neighborly Relations

"I have had some problems with neighbors complaining, and those who want me to tidy up. As I see it they just want everything to be in rows which are so boring to me."

Taking the high road, Jenny doesn't often discuss her garden style with her neighbors, suppressing her feisty-when-necessary spirit and letting them decide for themselves if they like it or not, wishing them well as they figure out how handle it however they can.

"It's the way I like things. Some people would like to see a different kind of garden and I say, well if that's what you like then why not do it in your yard? But this is mine and I'm going to plant it the way I see it."

Admitting that she is accustomed to occasional plant theft, Jenny places her more valuable specimens far enough away that someone would have to be really good to steal them.

"Taking bits of plants from here and there is part of being a gardener, that's what we do. We find things . . . but I have a person in my life who watches me like a hawk because he swears that I'm going to end up in jail somewhere."

When I commented that we might end up in the same jail, with a big smile and clap of her hands she giggled, "Oh, I would love it if they'd let folks like us have our own little gardens in the jail! And I wouldn't have to wash or do dishes!"

Social Acceptance

One morning in 2017 my neighborhood's social media app blew up with an alert that a neighbor had anonymously reported Jenny's lush garden to a city inspector.

The immediate reaction overwhelmingly backed Jenny, due in part to her friendly, nonconfrontational demeanor and partly to the diversity-loving neighborhood's live-and-let-live attitude. The online comments starkly illustrated the range of opinion, from "It's fabulous" and "Love seeing the butterflies and hummingbirds" to "Not my cup of tea." One person wrote "I love that garden yard. It was one of the factors that helped me feel good about purchasing a home on the same street. The citation makes me sad."

My favorite? "If she needs a lawyer pro bono, message me. Will work for cuttings."

The outcome of this torrent of support was a tactical move by the city inspector, whose reprieve basically noted that it was obviously done purposefully and maintained regularly—not simply a neglected garden gone crazy.

C'est la vie!

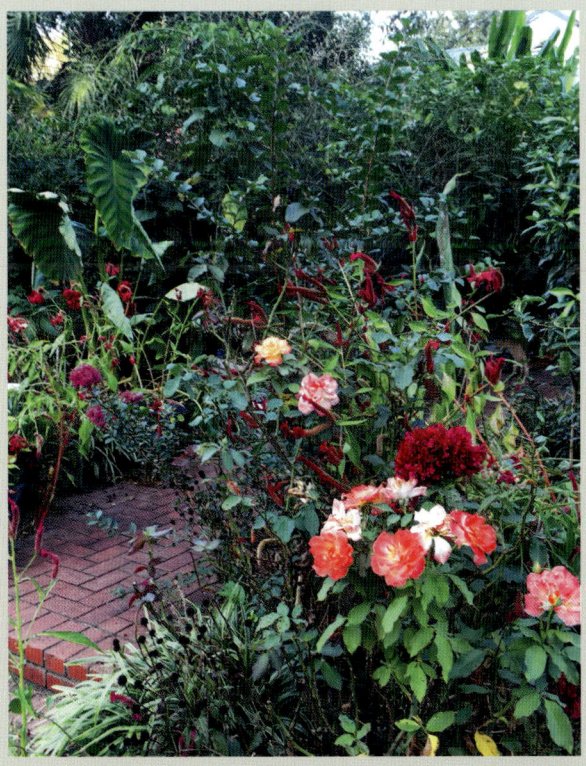

"But some of my plants I have a lot of, and people can take them. Besides, in my jungle if something goes missing or dies it isn't so obvious. And there is always something else to throw into the circle."

Beyond Beauty

In a seemingly common thread of spirit shared by such gardeners, Jenny doesn't grow her flowers simply for personal pleasure—it's an outreach of heart and mind. Her sharing spirit has her putting more than just plants onto the curb for people to avail themselves.

As a child in Jamaica, she and her family shared a cottage where they all gardened. "My aunt prepared every spring for an event called the Denbigh Show [the oldest and largest agricultural show throughout the Caribbean Islands]. It had a section for gardening, with prizes, and for weeks my aunt would be really excited about preparing her exhibits.

"I grew up poor in Jamaica, but very wealthy in all the things that are not measured as part of wealth. I didn't realize it at the time, but we were extraordinarily fortunate compared with many of our people. My family has always had this attitude of good fortune and we were raised, in a strong service attitude, to pass it along. We even grew flowers for decking out the altar and doorways of our church every Sunday, especially when the bishop came.

"So I don't garden just for myself. I try to plant the seed of this beauty with my neighbors. I give away a lot of plants, share what I can, and that is a big part of who my family is.

And it spills over. "When I worked as a nurse I always made sure our office had plenty of fresh flowers.

"That's why I will never move from this last home, I will not leave this busy main street, because the garden is there and it is a kind of ministry to the community."

Yeah, Jenny. *So nice to be different*!

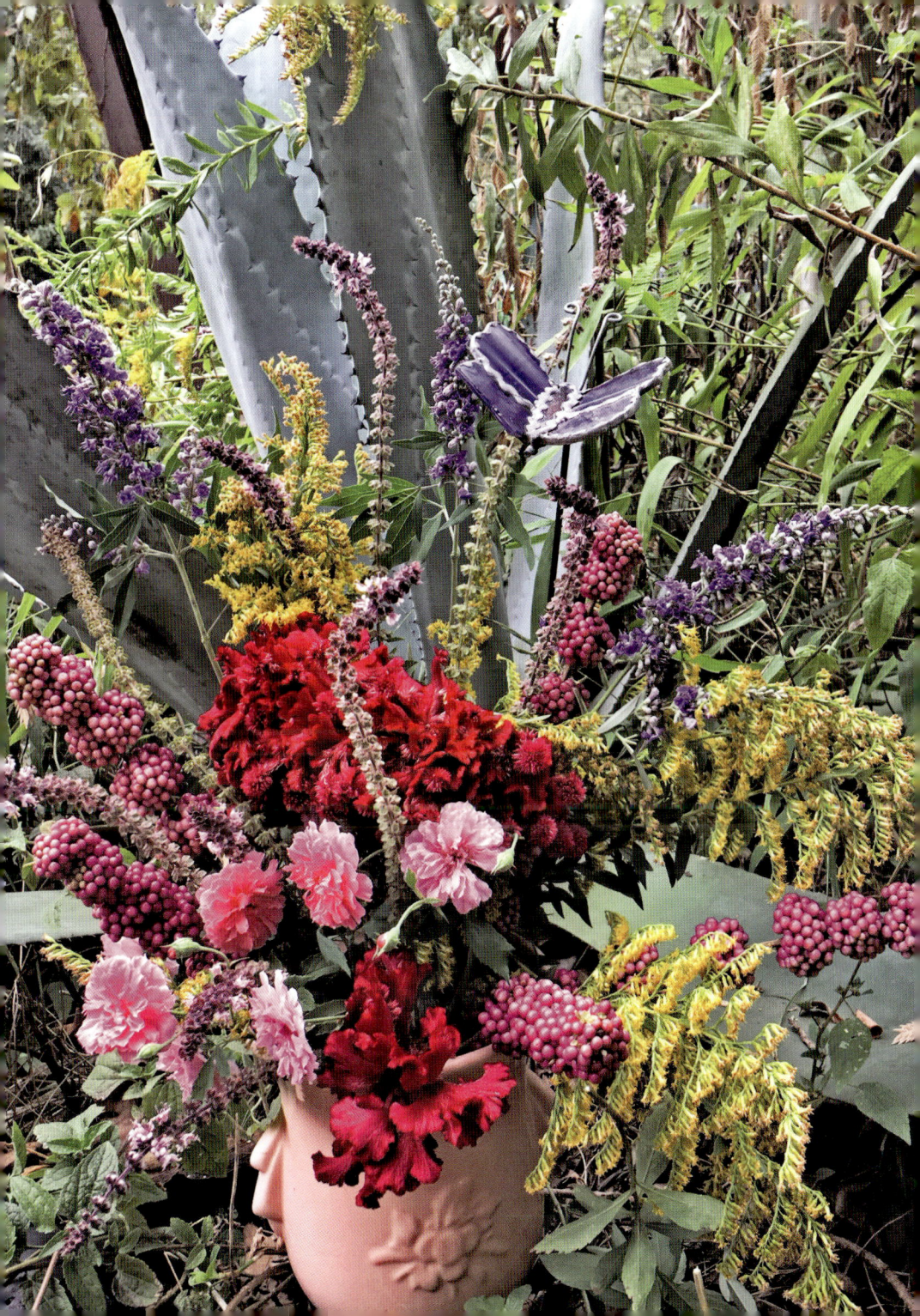

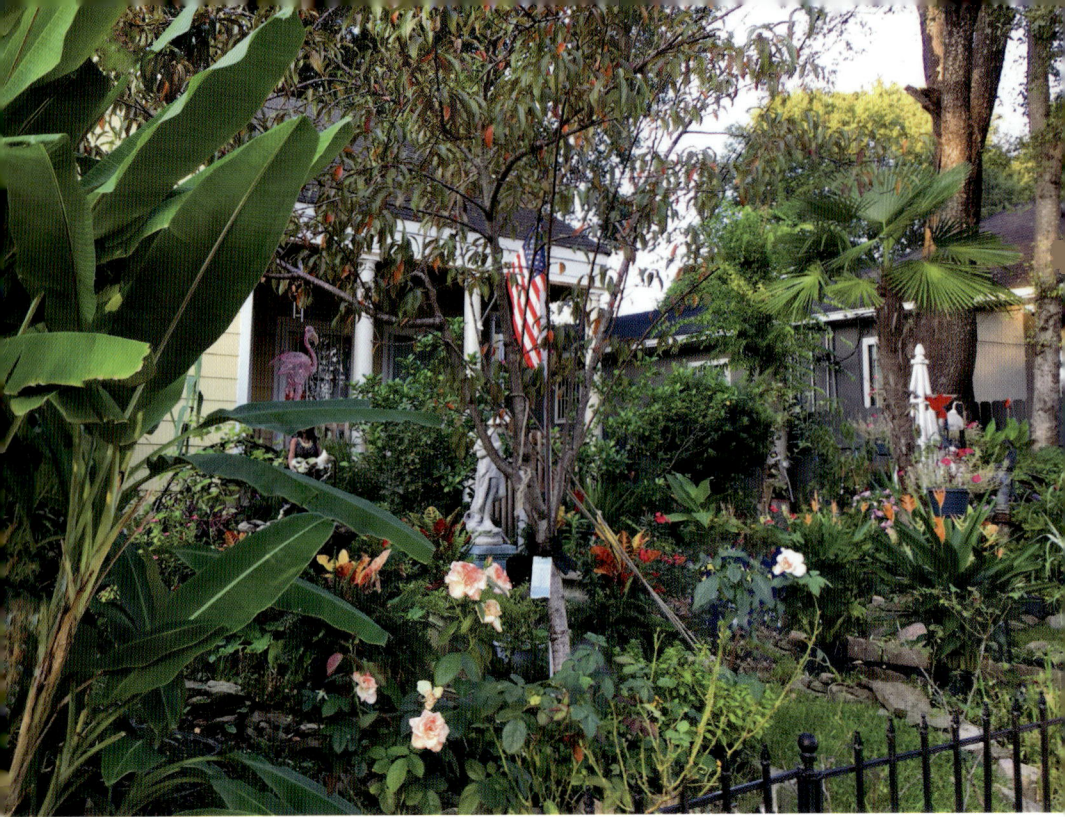

Perry Davis, the Evangelist

"I'm not supposed to be this way."

That's what Perry Davis, a big man with a bigger smile, sheepishly admitted one sultry Mississippi evening, laughing in amazement at his love of gardening. We were winding through his little fenced jewel-box garden, where, from my first visit, I felt profoundly peaceful and cozy, as if I had been born and raised there myself.

The small, neatly painted bungalow with its welcoming porch was once deeply shaded with little growing in the hard-packed clay beneath the old oaks. "I cut down the trees to let in some beautiful sunshine," he explained, "and just kept adding leaves to fluff up the dirt until I got it right."

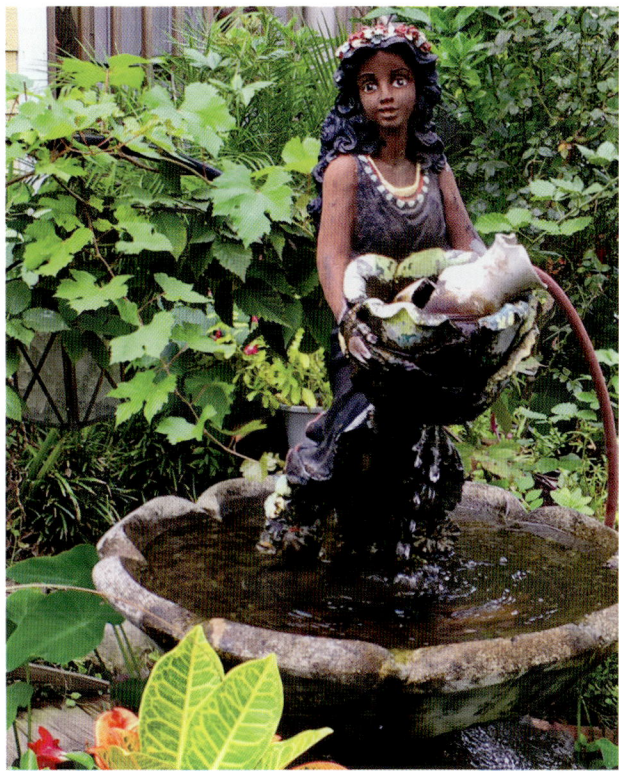

Now it's a meandering assortment of loosely packed groupings of small fruit trees, flowering and berry shrubs, towering bananas and palms, heady-fragrant gardenias rescued from construction sites, countless passalong annuals and perennials, and potted plants, all nestled around small water fountains, statues, figurines, and other yard art. The steps and walks are painted in bold colors; his concrete chicken is identical to the one I got from my own grandmother, only more gaily painted.

Perry has a natural knack for mixing shapes, textures, and colors, with repeated eruptions of plants with super-bold foliage including cannas, bananas, elephant ears, philodendrons, giant reed grass, tall ginger, orange-flowered "lobster claw" heliconia, and scattered pots of brilliantly colored African crotons. Among his big roses, butterfly-laden lantana, and other gaudy flowers, one of the most eye-catching fruit trees by his ornate front fence is a seedling from the burgundy-foliage 'Indian Blood' cling peach that Perry grew up with. He says it catches attention from everyone who passes by, including the postal carrier who commented on its beautiful spring flowers and sweet, fragrant fruits.

Perry's dad was a preacher who supplemented his income by working at a local garden center, and would occasionally bring home bits and pieces of plants, and taught his children to grow food in a huge vegetable garden. "Dad would take extra squash and corn around town to give away, but it was us nine kids who did most of the work. Each of us had our own area in the garden to tend.

"I was a bad kid, and mom made me do extra work. She whipped me with gardening, but it taught me to dig and fluff the dirt and plant things. And I guess I learned a lot, because I went from hating to plant stuff as a kid, to loving it now as art."

For several years Perry lived two houses up from Jenny Nelson, who nurtured his curiosity by bringing him plants. "She convinced me that it was better to have a garden instead of cars parked in the muddy front lawn. She opened my eyes, and showed me how to grow things the easy ways. It was nothing like I was raised doing in that hot vegetable patch; when I started getting into it, I had to go think back to Mom's memory to learn how to grow them. But Jenny sorta got me in a little competition thing, and shared plants and wisdom with me."

"I wasn't supposed to like gardening," he continued. "I used to think it was just little old ladies who did this but now I'm what we call a 'roony-poot'—don't have a college degree in growing plants but know how to garden from old times. Even if I don't know all the names, I do know that every flower means something to people who

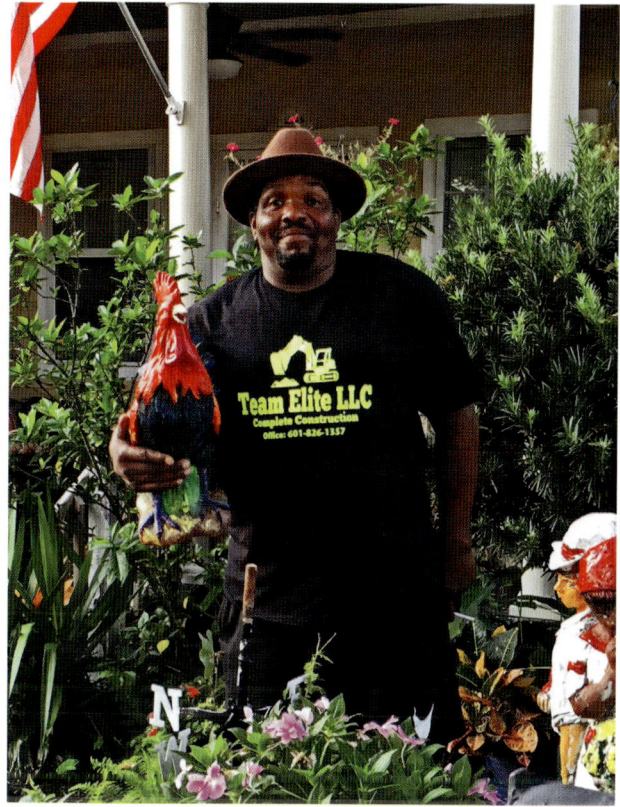

pass by. So I root cuttings to share what I can with anyone who wants a start of something.

"People don't always like the color of my house, or how I paint things up, but they love the spirit of my place. I see it change people's hearts when they walk by and stop to look, then smell, then start asking about the flowers. That lady across the street was happy when I started fixing up the place, and one day she cried when she saw her first hummingbird in my garden, feeding on flowers instead of manmade juice. She thought they were just a story."

Perry takes his faith seriously, saying the garden is like the Resurrection, how things come back better than ever. And it influences his belief in gardening as a social force. "We have to teach children. Some kids never see butterflies except in a museum, or hear the sweet songs of frogs in the evening. I don't know how that will change, but maybe we can help, I guess, one garden, one community at a time. But you gotta share."

Last time I chatted with Perry I commented on how though we garden as individuals, spread apart like we are, our flowers and gardens loosely connect us. He laughed and said, "Yeah. We're neighbors in dirt."

Joni Thaggard, Treetop Hideaway

To other residents of the top floor of Fondren's first suburban skyscraper, the only hint that something different blossoms behind the walls of their shared hallway is a shovel leaning against a door frame.

But those who make it into the eclectic apartment find an outer door leading to an astonishing pie in the sky, a delightful botanical smorgasbord spilling from large pots and vining up from lined beds set atop the building's roof.

It has all been created, bit by hand-carried bit, by Joni Thaggard, a transplanted small-town gal trying to fashion a retreat from the hustle and bustle of city life, a refuge from her stressful frontline medical profession where she can relax and sort worries.

Brushing through the lush growth, visitors find it hard to wrap their heads around how every pot, countless bags of soil and mulch, and the jungleful of trees, shrubs, perennials, annuals, and vines—not to mention furniture, accessories, and décor—had been hauled through the building's sole elevator or lifted hand-over-hand by friends, five stories from street level, with ropes.

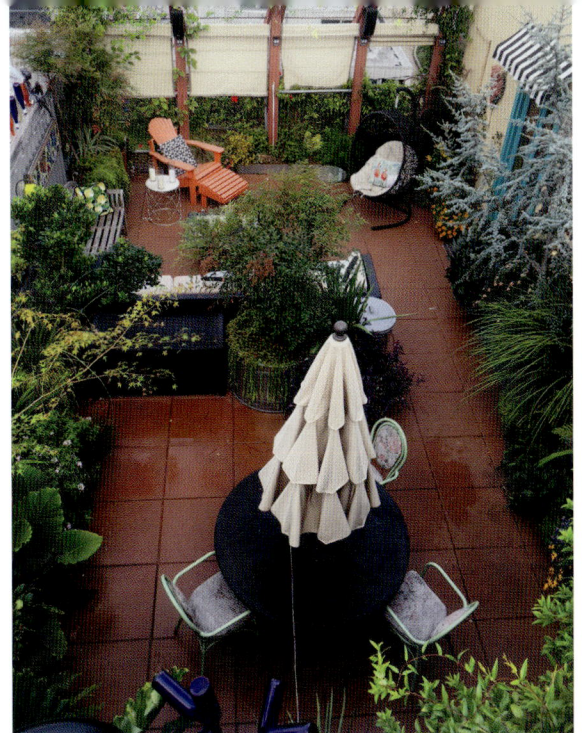

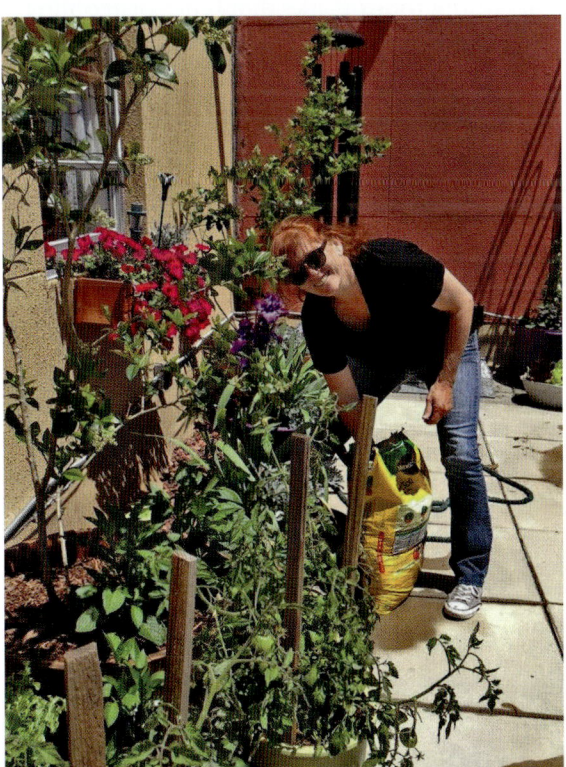

Credit: Josh Hailey

Joni, raised by pragmatic Great Depression survivors (who she describes as "minimalist bare lightbulb people"), admits that she wasn't sure how to get started. "I still don't have a plan *per se*, no overall design ideas. I just started filling pots with dirt and plants, one at a time, with I guess an innate sense of how to put things together to look good.

"I'm a plant freak," Joni sighs, acknowledging that she can't resist impulse buys and other-garden rescues. "I see a bare spot, figure out its conditions—the sun, the wind, radiated heat from the walls—and go out foraging for plants that might do okay.

"And I always come home with extras. I put them and little accessories together on a whim, and if something doesn't make it, isn't working well, there's no emotional attachment. I don't grieve if something's gotta go; I love my old cat, but plants can always be replaced.

"And you know what? The chance survivors—including some that experts woulda told me couldn't possibly survive on a hot, windy Mississippi rooftop, like that blue spruce and those maidenhair ferns and giant elephant ears—made me brave.

"Made me brave," she repeated softly to herself.

Eventually the sky-high eruption of plants began creating self-serving micro-niches of shade, shared humidity, and protection from the ever-present breezes. Ever the pragmatist, where live plants simply couldn't take the wind and heat, Joni carefully wove realistic artificial vines and foliage for year-round effect without needing watering.

When another tall building went up catty-corner across the street, Joni screened it out with a row of arbor supports, curtains and shutters which help deaden street sounds far below and can be opened for sunsets or closed for total privacy.

The once-bare patio is now divided with furniture and large potted plants into two rooms, one for dining and the other for entertaining around a gas-fueled fire table and watching sunsets over the treetops below. Tree limbs dripping with vines make the walls disappear, and

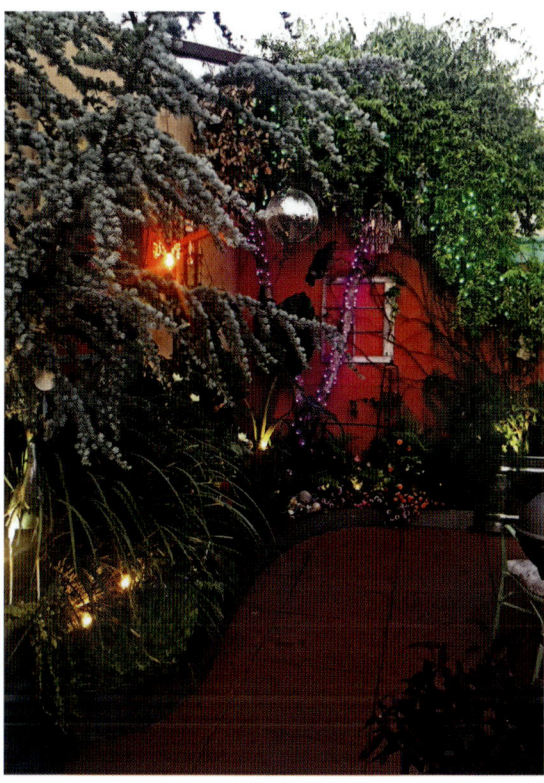

Magical faery den at night.

colorful glass bottles line the top of a heavily-decorated wall, glinting in what sunlight doesn't quite make it into the garden. Subtle lighting, both overhead and strung throughout the garden, provide soft ambience; reflective orbs add a Tinker Bell's faery touch and throw good feng shui into dark corners.

Explaining that her hidden garden is the only place where her innermost feelings are secure, she says "It's more mine than anything else in my life. It takes me back to my childhood, where I can almost feel the cool dirt on my knees from eating my way out of my grandmother's strawberry patch."

Soooo . . . what's the takeaway?

"The most important advice I can share for people wanting to feel right in their own garden is *just do you.* I don't quite have it yet. Not ever finished, can't put my finger on my obsession.

"But when I can't sleep, up at four in the morning, deadheading flowers in the soft, enchanted light is my therapy."

Jesse Lee Yancy, Guerrilla Gallimaufrian

I call him Sir Yancy; he comes back, in a lighthearted reference to his north Mississippi upbringing, with "Earl of Calhoun County, Knight of the Linoleum Table." But we both agree, for how he has transformed an empty urban abandonment into a both beloved and maligned oasis, that Jesse Lee Yancy III is the King of the Corner.

In spite of his self-deprecating demeanor, Jesse is not averse to taking on the entire world, starting in his own somewhat conservative neighborhood. "Some DIGrs may be 'crazy cat lady' gardeners," he asserts; "I feel more like a garage band, trying to shake things up."

In 2008 Jesse cautiously started a small flower and vegetable garden on neglected property across from his apartment. On land that he didn't own. Someone else's property. The owners of the space can take a mower to it, any time.

But for years, nothing had been done in the weedy, five- or six-foot wide space baking between the street and a little-used, shaded parking lot. With a "better to beg forgiveness than ask permission" shrug, he stood up a gnomon-less sundial and dug a few flowers, vegetables, and culinary herbs into the hard clay.

We crossed paths soon afterwards, following several of his somewhat pointed emails about my thoughts on the legality of what he was doing (including planting cotton other than on a real farm, forbidden by an obscure agricultural regulation).

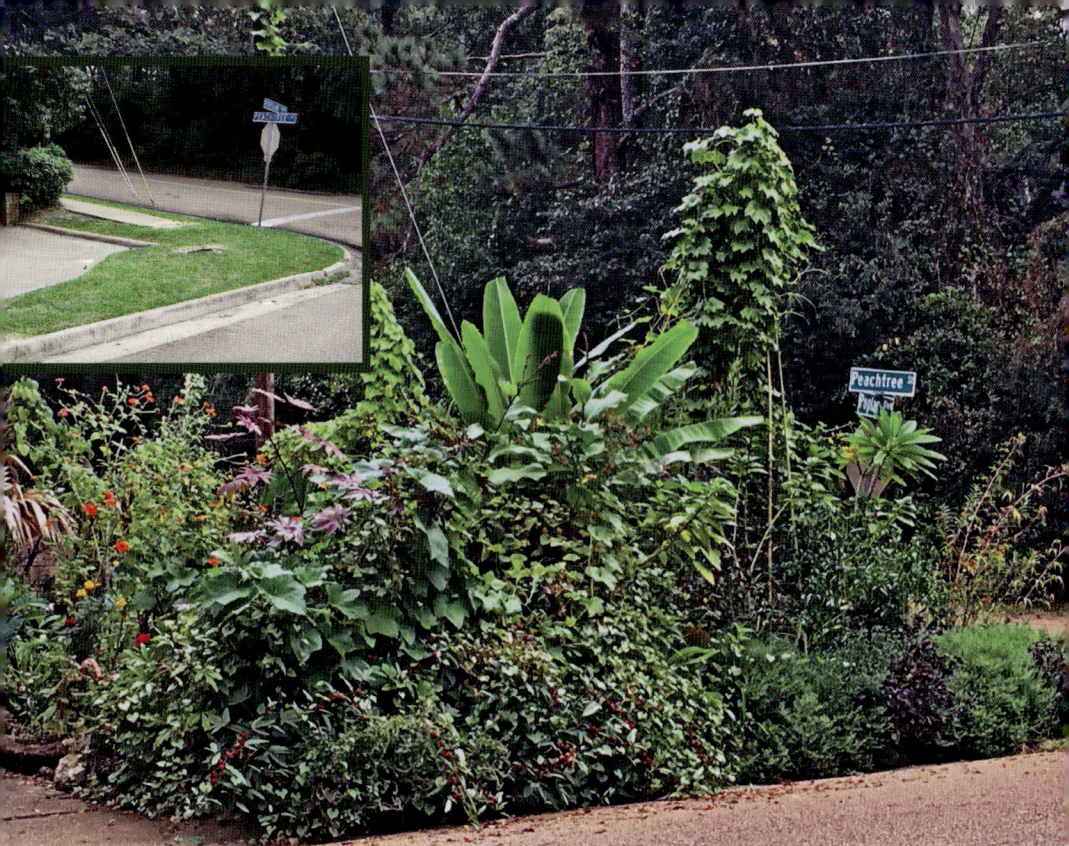

Turns out, Jesse, whose conversational face belies a fierce advocacy spirit from social issues to saving trees at a local park, is not your run-of-the-mill dabbler. He invests physical, mental, and spiritual effort into his insecure garden, just as he brings his university literature background to virtual pen in his prolific, lyrical blog about . . . well, everything Southern, including beloved writers, classic Southern cuisine, heirloom plants, social relations, and local history and lore.

The garden he developed quickly became a mulched horticopia of Southern heirloom plants and cast-off broken *objets trouvés*.

From the street, as with most DIGr gardens, it appears a chaotic tumble of flowers, vegetables, herbs, and tropical plants, patches of seedlings tucked between withered wildflowers whose seed are drying for next year, vine-covered arches and trellises, assorted containers including buckets and inverted tires, unkempt piles of soil, compost, and mulch, and rough little walkways winding through it all.

However, after just a short chat with its creator, its long-view sensibilities emerge. "My little corner of the world is, as one person put it, a 'garden of the moment' as if there were such a thing. And while I've learned a lot from other gardeners, most of the best lessons I've discovered the hard way, by screwing up and having to correct them.

"After the death of my last remaining sibling left me at loose ends, I started the garden as a form of therapy more than anything. Over time it has helped me regain focus—gardening is a patient art, and it makes you slow down and look at things. It also helps you learn how to care, to think outside yourself.

"The garden grew slowly, and it's probably better that I don't have a truck or equipment because that has taught me to use what I can find: fallen leaves, sticks, pieces of broken concrete, discarded lumber and wire. I work with what the world provides.

Hügelkultur

When it came to choices of where to grow plants, Jessie had three: containers, dense Yazoo clay, or raised beds atop concrete paving.

His pots are filled with any potting soil he can get help hauling. The hard clay is hand-dug as deep as practical, the clods broken up and mixed with leaves, bark, and compost. With each digging, planting gets easier.

But atop the concrete of the parking lot, he shaped beds with logs and tree limbs, and filled in with more limbs, branches, leaves, and whatever else he could glean, topped with compost. It's an ancient practice called *hügelkultur*—mound culture, which is ideal for difficult or dry sites; once planted, these materials break down slowly while holding moisture and releasing nutrients, eventually becoming decent soil. Takes time at first but, as he puts it, "It's a lot like cooking, starting with potatoes and gravy and building on that."

"I'm not a GREAT cook," Jesse demurs. "Worked as a journeyman chef for fourteen years but I'm not one of these geniuses you read about in the foodie press. I do, however, know what will work and won't work in most any given situation.

"Same with my garden, unsophisticated with its pell-mell of plants that will grow hither and thither. Having said that, I am proud of its success in having even become a garden in the first place. And though my resources are quite limited, I'm very proud of my little *pied a terre*.

The Plants

From midwinter antique daffodils to late autumn asters, Jesse grows an astounding menagerie of unusual plants. Black castor bean and brown cotton lock in a season-long *pas de deux* amidst the swirling ballroom of burgundy okra, bright red roselle, edible greens—mustards, turnips, Brussels sprouts, kale, collards, and colorful lettuces—and all-season wildflowers. He also provides a safe refuge for faded poinsettias, Easter lilies, slumping jack-o'-lantern pumpkins, and other cast-off holiday plants, creating seasonal hedges with them.

A good gardener has to be sensitive to the slightest differences in conditions. And, tiny as the garden is, Jesse knows exactly where the sunny areas stay moist longer than others, how much shade is acceptable for sun plants and how much sun the shade-loving plants can tolerate.

"I've discovered to start big annuals like cosmos, peppers, and sunflowers in small containers and transplant rather than scatter-sowing and thinning. It gives the spring flowers time to bloom out and give up some room.

"And when the cold comes, I cover cardoon and fledgling hollyhocks, since they're in the path of the rolling frost that flows down Peachtree Street and leaps over the hill into my garden."

To protect his many potsful of begonias, aloes, ferns, a huge plumeria tree, and other frost-tender tropical plants and succulents,

You So Country!

Jesse Lee and I have had misunderstandings and dust-ups, often over trivial matters such as how he calls the small, golden, very fragrant wild gourds that spring up plum-grannies; I call them pocket melons. Whatever.

On one visit I openly admired his straightforward but spirited "make do" approach to gardening, laughingly calling him "so country" which to me is a high compliment. But Jesse's pride took umbrage, assuming that I was looking down my horticultural nose at his simple methods.

We salved the chaffed feelings over a potful of my great-grandmother's bulbs, after I confessed that I have long gotten the same guff from others about my own garden—I, too, was taught by country women to nurture a mishmash of favorite plants and love over-accessorizing.

But c'mon, Jesse—you once told me you learned to "fish" for doodlebugs with long grass stems in the fifth grade from the only girl you ever met who knew how to whistle between her teeth. If that ain't small-town country, what is?

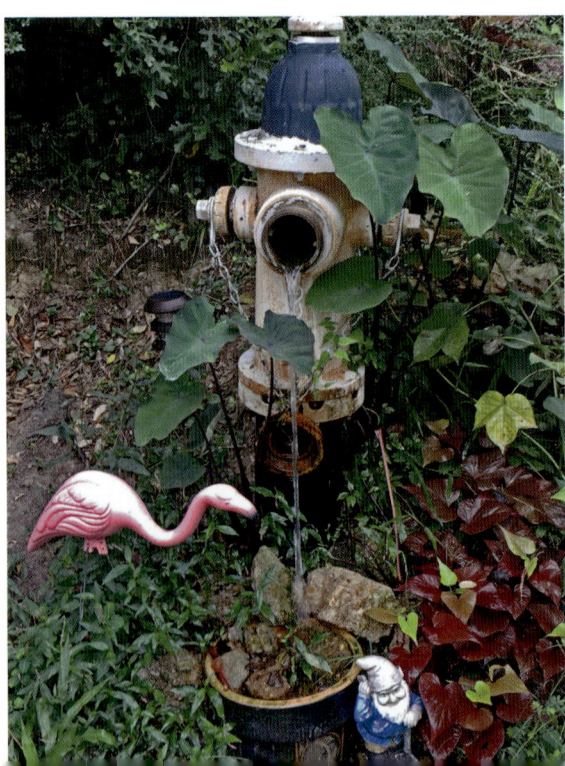

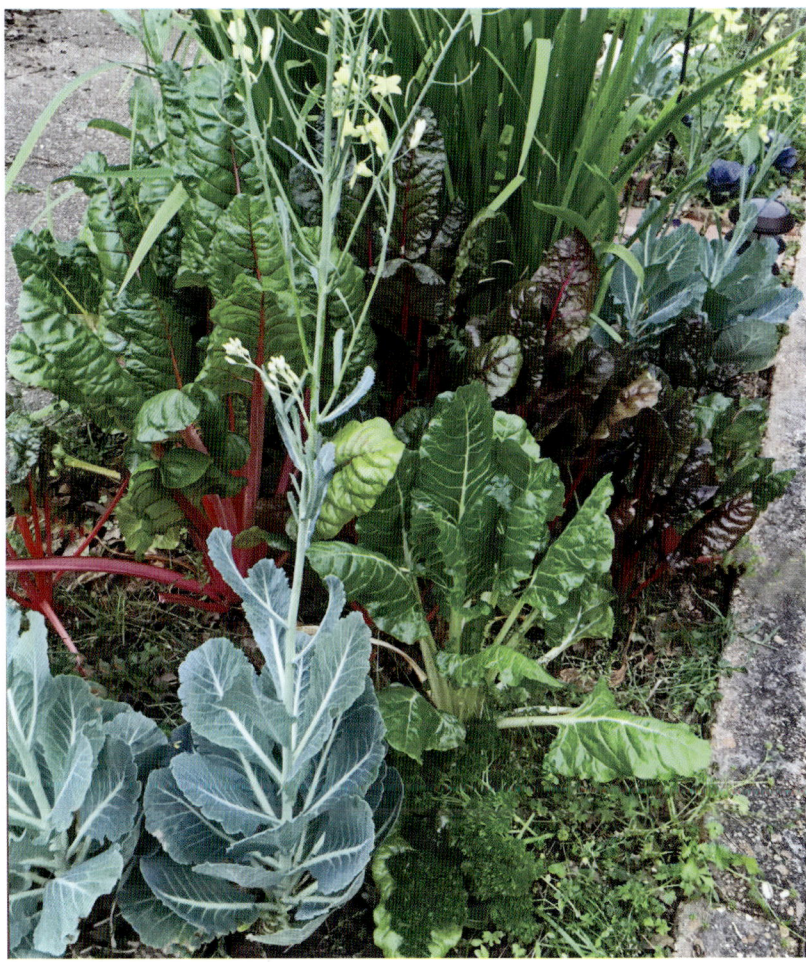

Jesse trots them in and out of a storage room in the basement of his apartment across the street.

"I've come to the belated conclusion that *there's no hurry*. I mean, good grief, we have a nine-month growing season here, and I've finally stocked the space with enough pretty perennials (however

Ark of Taste

Jesse's garden is an informal corner-of-the-world test plot for what the International Slow Food Foundation calls the Ark of Taste which collects and celebrates the sometimes-obscure food plants that help define cultures.

Jesse showcases and shares some of these prized culinary rarities as a way of helping stem the ebbing away of the extraordinary traditions of which they are part.

To highlight just one, there's an unknown garlic he calls Pocahontas. "When a friend from Pocahontas dropped off his garlic for my garden all these many years ago, he piled the dried knobby stems in a haybale near the parking lot wall, and ever since then I've had Pocahontas garlic coming up there. In the late winter the flowers, all lovely to behold, nod like old men in a spring sun. It's a tough plant, always late no matter where you plant it (at least it is for me) but keeps going and is prolific."

run-of-the-mill they might be) not to have to worry about getting the annuals in when the daffodils bloom.

"It's a constant struggle, finding room for everything and making room for new. Not to put too fine a point on it, anything that's in my bed for six months and doesn't put out simply has to go!"

Jesse sometimes goes onto our neighborhood's social media site to share surplus plants and vegetables. The responses pour in quickly, with counter-swap offers of dried herbs, home-canned chow-chow, or strawberry plants and the like.

"My corner garden is very much a passalong garden, not only because I don't have a lot of money but also because the garden was designed from the beginning to be a "mother ship" for neighboring gardens.

"People can pass by and browse, and they share plants and seeds. Most times this simple act of sharing is the beginning of a friendship, and more often than not the friendships last longer than the plants.

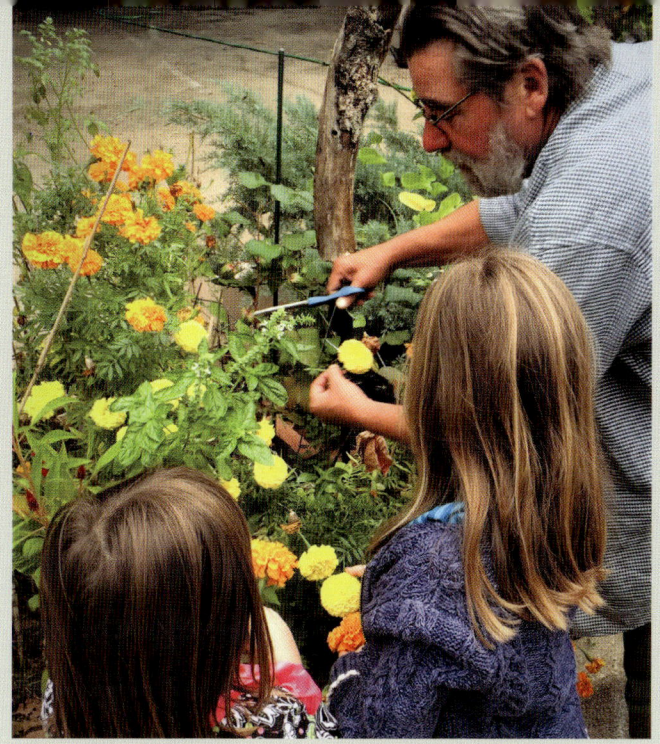

Jesse encourages through sharing.

Little Corner Herbary

Jesse shares much of his largesse with neighbors, helping newbies get started, and donating extras to local plant sales. But with a determined nod towards keeping on the good side of everyone, he takes it a step farther.

Jesse's community has informal libraries—colorful weatherproof stands where neighbors freely drop off and borrow books from one another. But for years Jesse has pioneered the "little corner herbary" concept in which he carefully places culinary herbs where neighbors can snip a little rosemary, oregano, or whatever they need.

"Height and color are primary visual objectives when it comes to street traffic, but scents, and something good to eat, can quickly pull pedestrians in and get them hooked. Especially children.

"It's not entirely altruistic, just to keep these plants and practices alive with new people; if I don't make the corner a neighborhood resource then there's every chance of losing it to someone's vapid idea of a "neighborhood improvement" project."

"The more people touch and eat from my garden, the more they learn to love it."

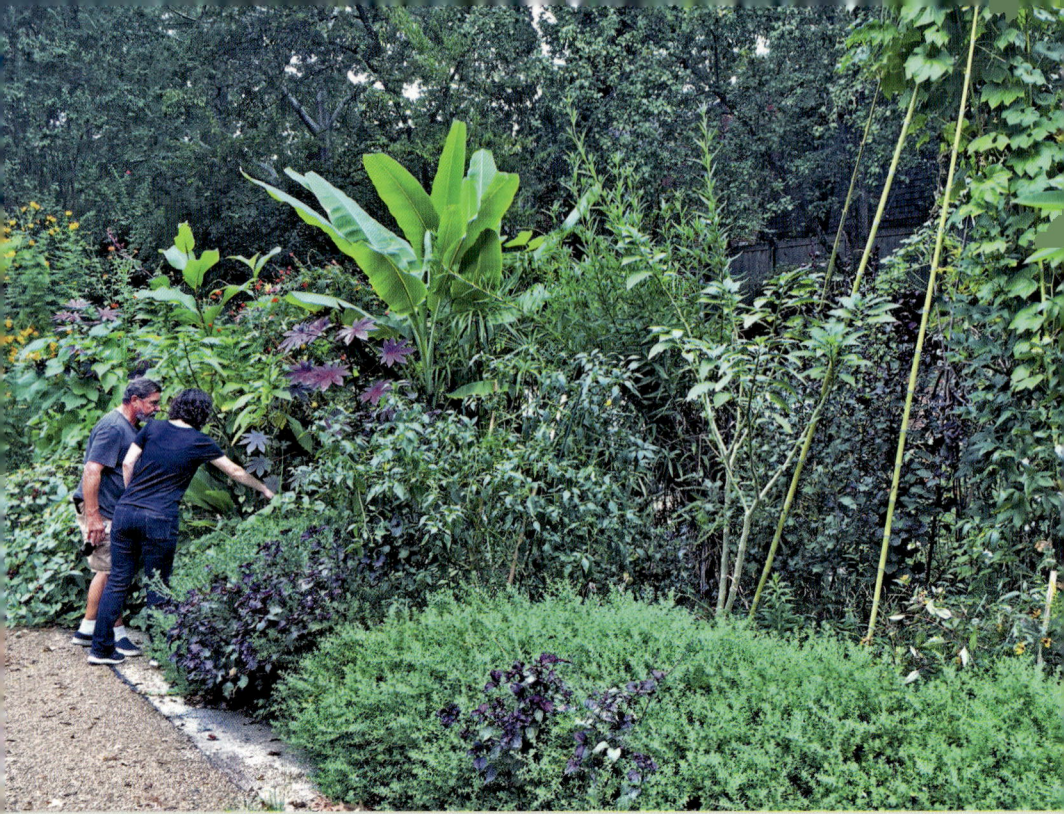

On Being Watched

"I come from a small town in north Mississippi where people are habitually friendly and cordial. But here in the city, people walking their dogs or strolling with kids might not greet me while I'm puttering in the garden, which I find very disturbing.

"A few will stop and chat and maybe tarry a bit to watch me digging, weeding, or pruning. But others just stand there and stare at me or even discuss what I'm doing between themselves as if I were some sort of deaf automaton. I find this very strange; am I crazy?

"A lot of earnest folks who come by give advice, and I'm grateful for sure; but sometimes I have to just smile and feign to agree, then go on with what I was doing.

"There are lots of kind neighbors who take me to get mulch or help me move heavy things. They also give me art to put in place, gnomes and pretty rocks and old trellises and all sorts of sundry things. I've had to find a place for everything, because they're going to come looking for whatever they gave you one day."

Growing Along with the Seasons

Jesse shared how Roger Swain, longtime host of *The Victory Garden*, once proclaimed that Mississippi IS a garden. "And he was right. The problem is, it's not being cared for. I simply found a piece and started caring for it.

"It wasn't my piece of Mississippi, but that's what guerilla gardening is all about: gardening on someone else's property. If the city ever decides to rework the neighborhood street, my garden might be paved. It's been fun, but *sic transit gloria mundi* (thus passes the glory of the world).

"I've come to understand and embrace the corner garden as a whole rather than a series of beds along an angular stretch of asphalt. I'm determined to let things run their course, grow and flourish as they will and should, and I've already concocted new projects that will make use of most of it.

"Meanwhile, I'm conquering what I can one foot at a time, and things are growing apace. It's been a lot of effort; my old body is displaying previously unknown aches but frankly I feel the better for it all, as evidence of physical competency if nothing else."

Plants: Growing It Forward

> Congratulations! Today is your day. You're off to Great Places! You're off and away!
> . . . And will you succeed? You will, indeed!
> —Dr. Seuss, *Oh, the Places You'll Go!*

DIGrs would be . . . *nothing* . . . without their plants, their little cultivated bits of *Yggdrasil*, the mythical Tree of Life. Much like food, music, and sports, shared plants are an unspoken language that unite an incredible diversity of people across space and time.

Compared with frozen TV dinners, they are potluck buffets, plates piled high with comfort food.

Not mere floral pets or food-making workhorses, they often have meanings far beyond their most practical horticultural or alluring sensory uses. They testify to generations worth of beliefs and social and historic connections. Entwined with precious stories, they create a vast, informal web of people, places, and lore.

Driven by curiosity and an admiration of that which persists, over many years I have made it a point, wherever I've found myself, to photograph and compile lists of commonly grown regional passalong plants—the ones most likely used by DIGrs. These resources have proved incredibly handy for custom-crafting my presentations for different crowds, and I'm happy to share some of them here.

With some modifications due to climate and or soil differences, there is a lot of overlap of exceptional plants that are commonly grown all over the world, all grown pretty much the same way with minor modifications.

Still, wherever people garden they tend to cultivate combinations of locally adapted plants that not only survive, but also create a strong "sense of place" like no other region.

Gathering these plants, then moving them on down the line, is called *growing it forward.*

Heads up: Passalong plants and the passions they generate are quite infectious, a bit like cold germs which can spread rapidly from gardener to gardener.

Passalong Triarchy

Like a popular socialite, to make it as a widely planted heirloom a plant has to stand out from the crowd and yet be amiable with others.

It should meet some basic requirements, which I call the Passalong Triarchy. None are ironclad, but the more of these characteristics a plant has, the more people of all skills and interests will grow and share them.

First, a plant should be valuable. Common values include beauty (flowers or foliage or both), fragrance, reminiscence of something (family, home, culture, history), edibility, appeal to wildlife, whatever; the more values, the more gardeners will appreciate it.

Second, it should be easy to grow without a lot of expertise. It won't be enjoyed by a lot of people if it can't tolerate a wide range of growing conditions (soils, temperature, and water extremes), resist common plant insects or diseases, and not require routine pruning or other untoward maintenance.

Finally, no matter how valuable or easy to grow a plant may be, it won't go very far unless it is almost pathetically easy to propagate

---

Gardening through all seasons.

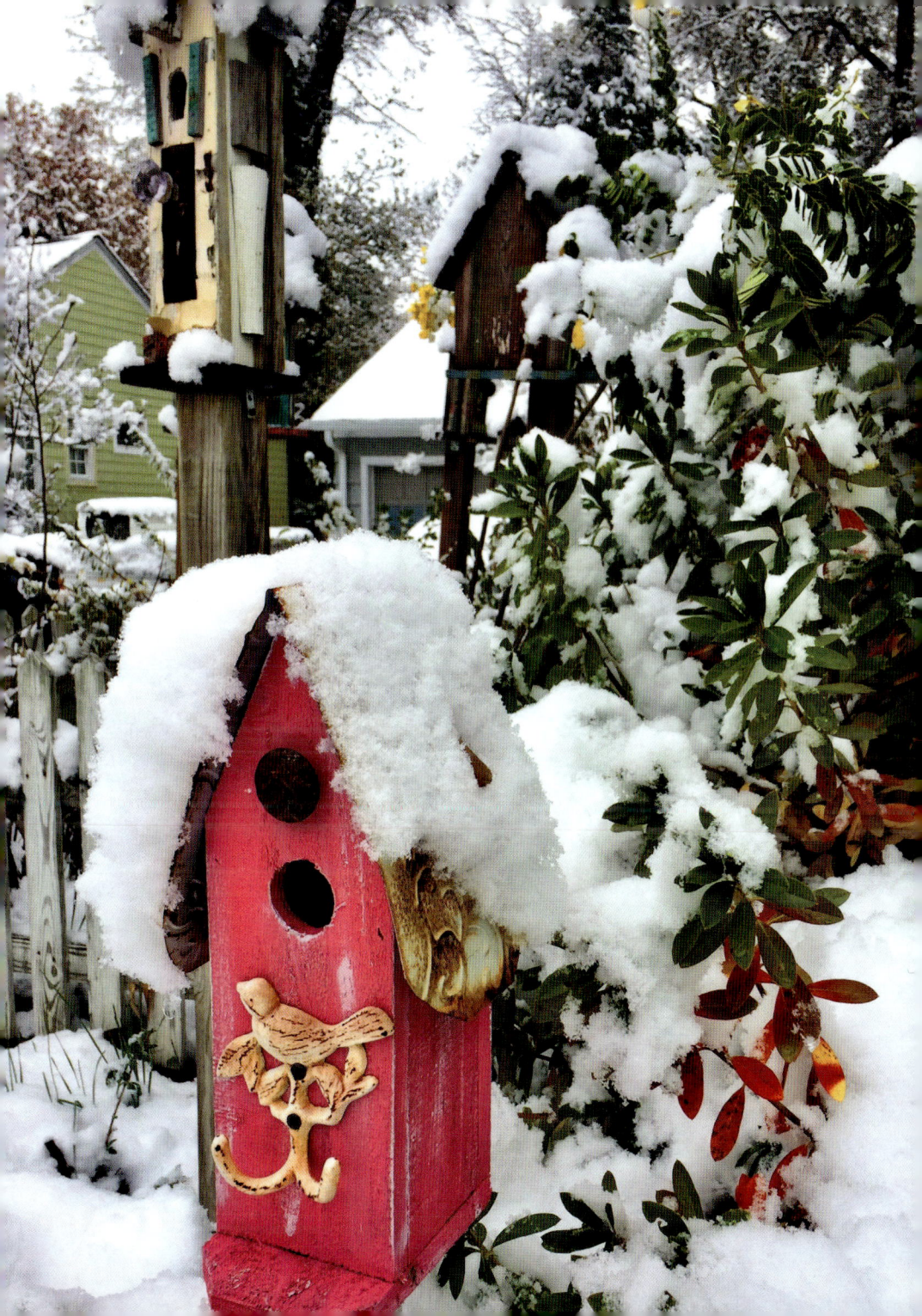

from seeds, divisions, or cuttings. The easier it is to share, the more it'll be shared.

Still, DIGrs love to push boundaries, and will often be caught growing plants that the experts will say normally don't have a chance. Can't tell you how long it took to me to learn to not publicly declare something is unsuited to be grown in a given area, because inevitably someone would come up to me afterwards to share condescendingly that "Mama has grown that all her life here."

World and Regional Favs

To keep things simple in this book, which is about attitude more than how-to, here are some of my uber-abbreviated lists of the most commonly shared types of plants I have seen in small gardens in various areas around the country and world.

Keep in mind that there are some glaring regional exceptions. Some garden plants are popular worldwide, but there are many unique species and even varieties or cultivars that are only hardy in some areas. A fern grown easily in Canada or Germany will probably not be the same species found in Florida or California.

That said, gardeners in Europe, the American Midwest, New England, Pacific Northwest, and most mountains are likely to share cold-hardy rhubarb, peony, hollyhocks, snowdrops, hellebore, and cranesbill geranium. Southeastern US and Mediterranean-climate gardeners will be more likely to share heat-loving figs, liriope, four o'clocks, aspidistra, purple heart (*Tradescantia pallida*), agave, ruellia, angel trumpet, elephant ears, paperwhite narcissus, lycoris, banana, and lantana.

In frost-free areas and in container collections all over, gardeners share various tropical plants such as banana, gingers, aloes, bromeliads, and tender succulents.

You know there are many dozens, even hundreds, of others, with a lot of overlap worldwide. But to illustrate how complex any DIGr garden can be, let's look at a few which represent the many that can be found anywhere on Earth.

Yeah, I dare say I left out a few of your favorites, but a casual walkabout in an older neighborhood, or visit to a nearby botanical garden, or certainly any plant swap will likely turn up jewels I've overlooked here.

On Plant Names

Tow-MAY-toe, tow-MAH-toe, 'erb or h-erb, whatever. *Please don't hold your nose* if I mispronounce or refer to a plant by an unfamiliar name or folk name. Comfortable gardeners often speak plainly to one another, slipping into relaxed country sayings and clichés that cause outsiders—and botanists—to shudder. Same with folk names for plants.

Heck, I can't find two gardeners who pronounce *liriope* the same.

It's not that we don't know better; though cognizant of the oft-confusing rules of language, sometimes we prefer less-precise downhome lingo over more polished, highfalutin' discourse. A lot of folks who can't bring themselves to communicate in a local jargon, or criticize those who do, are just putting on airs, bless their hearts.

So, in my effort to keep things simple, I am eschewing precise taxonomy (which changes, often), opting instead for the monikers most of us come across the most often. If you are unfamiliar with any, simply look them up online or in any decent garden book.

Most Common Passalongs Worldwide

Canna
Daylily
Euonymus
Ferns

Garlic
Iris
Mints
Oregano
Ornamental grasses
Pepper
Prickly pear cactus
Purple coneflower
Rose
Rosemary
Sedum
Tomato
Yarrow
Yucca

Tropical Potted or "Porch Plants" Favs

African violet
Airplane or spider plant
Aloe vera
Asparagus fern
Begonia
Bromeliad
Chinese evergreen (*Aglaonema*)
Christmas, Easter cactus
Crassulas (jade, etc.)
Dieffenbachia
Euphorbias (many including pencil cactus and crown of thorns)
Hen and chicks (*Graptopetalum* and *Sempervivum*)
Mother of thousands (*Bryophyllum* or *Kalanchoe*, others)
Night-blooming cereus (*Epiphyllum*)

Peace lily (*Spathiphyllum*)
Pelargonium geranium
Philodendrons (especially heartleaf vine)
Pothos vine
Ribbon dracaena
Sansevierias
Scented geranium
Strings of anything (pearls, hearts, etc.)
Tradescantia (Wandering Jew)
Walking iris (*Neomarica*)
Wax flower vine (*Hoya*)

Plants That Even Dead People Can Grow

The term "pushing up daisies" has a certain appeal beyond the graveyard.

Old cemeteries, which at one time were the earliest botanic gardens and public parks in America, were the pride of small towns and usually had something every season. People would go at least once a year to clean up the places and often hold family reunion picnics on the grounds afterwards. Some still do.

Many of the irises, daylilies, hostas, daffodils, trailing sedums, roses, and other shrubs and small trees are planted as meaningful and long-lasting memorials to those left behind, and often come poignantly from the gardens of the deceased.

All the plants that survive in these memory gardens are tough; they grow in just plain dirt, usually on rainfall alone, and don't require a lot of maintenance. And they must live for many decades, maybe even centuries; in fact, many actually outlive the grave markers.

Plants dead people can grow: Sounds like stuff that would suit any sustainable year-round home garden don't they?

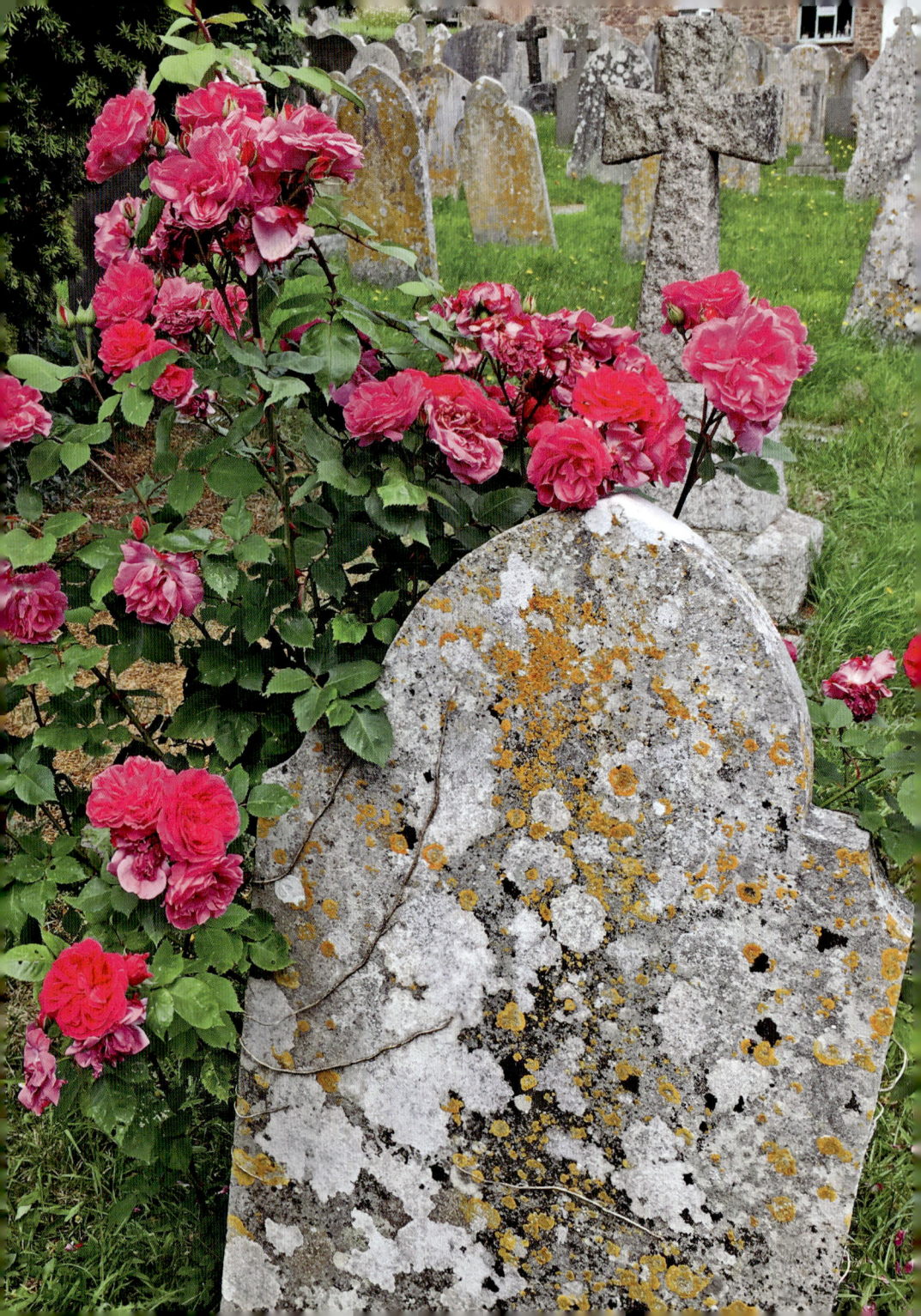

British Passalong Plants

Many DIGr gardens epitomize the overflowing fenced or hedged cottage-garden style we associate with English gardens, where quirky and overstuffed are inspirational terms. They give DIGrs credibility.

For over thirty years I have explored gardens in every corner of the British Isles, and have gardened for ten years in the north of England. Been a member of both the Royal Horticulture Society and Cottage Garden Society. Participated in the oldest-running plant swap in England, and have taken a particularly keen interest in heirloom plant displays at countless flower shows, both large and small.

And on year-round rural rambles with my walking club, I pass a lot of sweet country and small village gardens, noting appreciatively both native wildflowers and garden escapees along footpaths, some of which are banned outright for their aggressive behavior.

With enthusiastic help from various heritage plant groups and the organizers of PlantSwap UK (motto: "Bringing People and Plants Together"), I have honed a short list of classic British passalong plants. Many are also commonly shared in America's Pacific Northwest and other similarly moderately cool, Mediterranean-type climates.

Alchemilla mollis
Bergenia
Campanula
Columbine (*Aquilegia*)
Dianthus
Geranium (cranesbill)
Foxglove
Hellebore
Hollyhock
Jacobs ladder (*Polemonium*)
Japanese anemone
Lythrum (several)
Mint
Montbretia (orange Crocosmia—
　　top invasive passalong)
Penstemon
Pulmonaria
Rose campion
Rosemary
Saponaria
Saxifrages (including strawberry
　　geranium)
Sempervivum and other succulents
Strawberries and vine or cane berries
Verbena

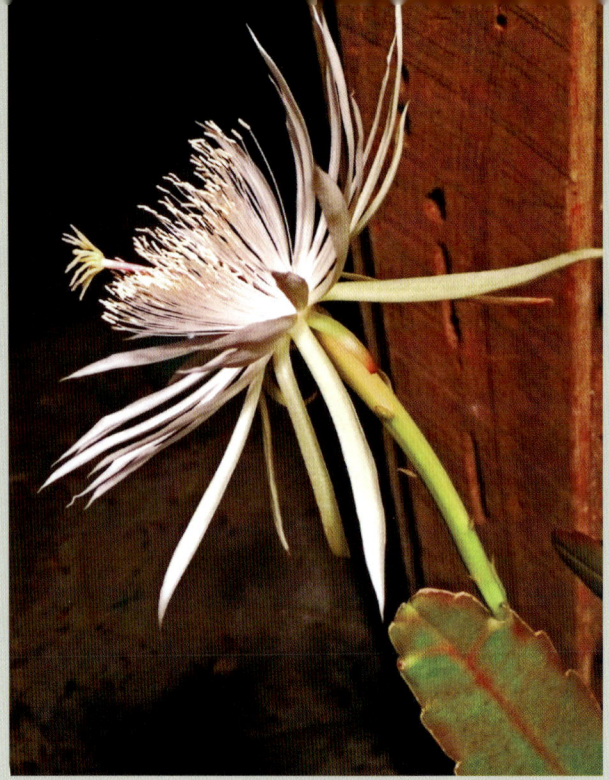

Plant Provenance

The best passalong plants come with a story of provenance in tow. When Dirt and I visited Tom Mann, a neighbor with an extraordinary collection of hardy fruit plants, we were both given cuttings of a tropical vine called "wax flower" (*Hoya carnosa*).

The vine had come from Tom's mom in the late '70s, who in the late 1940s had gotten it from her own mother-in-law, who had previously received her starter piece from her Jackson dentist who knows when?

This is similar to the hand-to-hand source of one of my great-grandmother's favorite old parlor plants called night blooming cereus (*Epiphyllum oxypetalum*) which I have shared with countless other gardeners. The story goes that she got her "start" from the president of her garden club, who had gotten hers from a friend who got his during a "flower watching party" held by celebrated Jackson author Eudora Welty.

I once mailed rooted cuttings of this scraggly old cactus to over a hundred supporters of my National Public Radio program, spreading it to who-knows-where. I've gotten photos emailed back of their success.

What have you got like this? Or is it time to start your own thread?

What's Native, Anyway?

Sorry, native plant buffs, but our gardens and cultural heritage are forever influenced by plants from all over this hugely diverse world, chosen for their enriching beauty and nutrition.

As past president of my state's native plant society, I deeply appreciate the benefits of using local plants both for us and the wildlife web.

However, what garden doesn't include plants from all five of the inhabited continents, including one or more that's likely to become rampant (as DIGrs put it, "Be careful, this'un will get away from you")?

It's easy to say what is native to the Americas—basically anything growing there before Europeans began bringing stuff from all over everywhere else. But in Europe and Asia, it's anything growing there since the Ice Ages.

Kudzu, English ivy, and other invasive exotic menaces in America's natural areas are on par with the American goldenrod and Himalayan balsam that have spread from proper English gardens; there are other love/hate escapees in Britain including montbretia, Spanish bluebells, and buddleia, all great garden plants but seriously invasive. I can go on and on, but you get it.

Main thing is, today's flower and food gardens wouldn't exist without the world's universally shared largesse. Yet, as Roger Swain put it, "Few gardens outlive their gardeners; but some of their plants outgrow the garden."

Be careful what you plant – it could become your neighbors' nightmare.

North American Natives favored in posh European Gardens

Amsonia
Coral honeysuckle
Coreopsis

Echinacea
Gaura
Goldenrod (many)
Helenium
Helianthus
Heuchera
Hymenocallis
Joe Pye weed
Lobelia
Louisiana iris
Monarda
Penstemon
Phlox (many)
Physostegia
Prickly pear cactus
Rudbeckia
Star hibiscus
Stokesia
Sumac
Tradescantia
Violets
Yellow evening primrose
Yucca filamentosa

Hedging Bets While Trying New Plants

There's something alluring about sampling the unknown.

Though I don't get my shorts in a knot over faddish new plants, neither am I stuck on the same old, same old. Truth is, some brand-new introductions are great, and may actually pan out to become favorite passalongs someday on their own merits.

American natives are mainstays in English gardens.

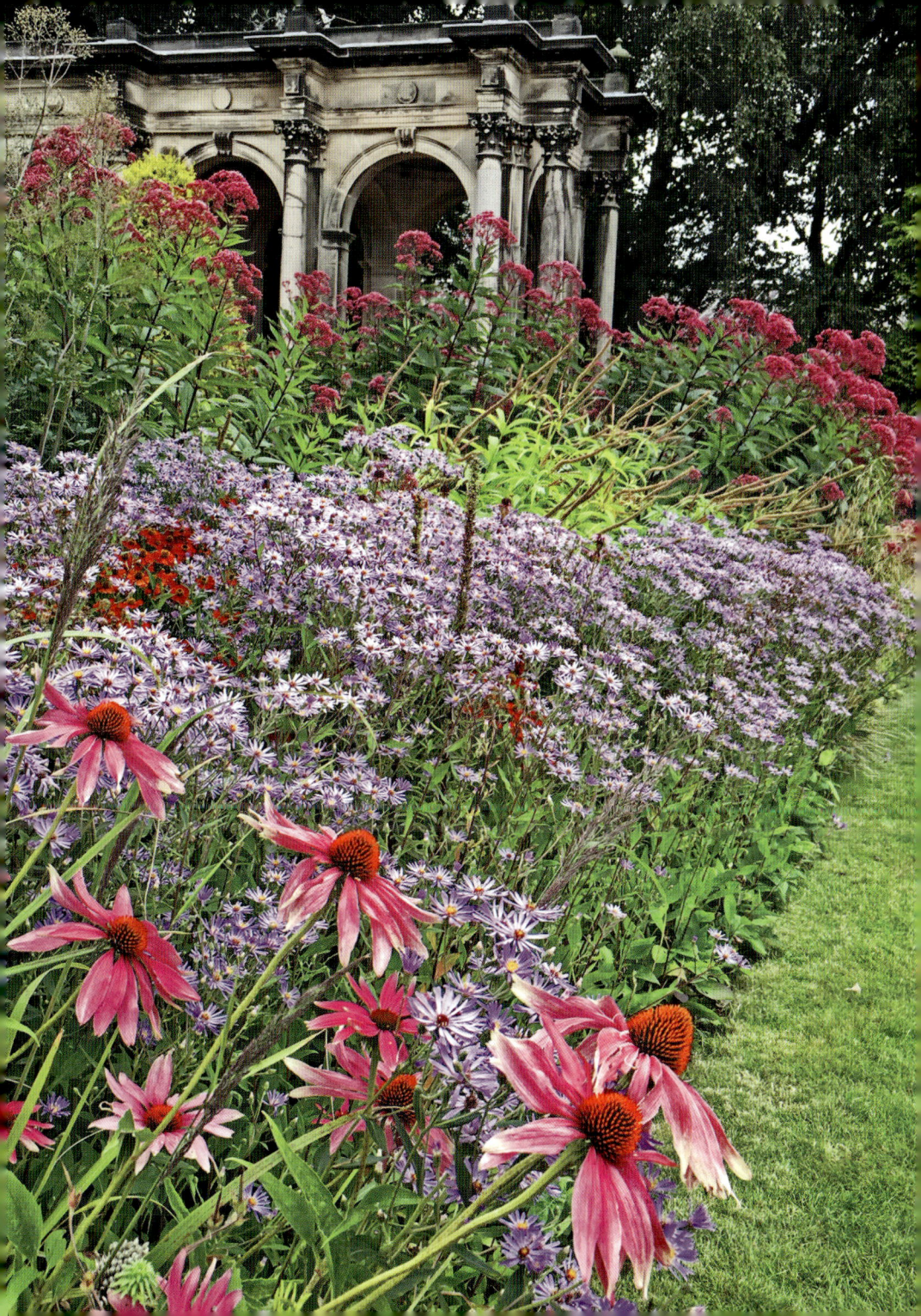

But there is a deep psychological reason we fall for new plants. It isn't just prettier flowers or better flavor, longer production with fewer pests—there's an evolutionary advantage in it as well.

Seeking new experiences is a fundamental behavioral tendency in both humans and animals, helping us try new things that may prove advantageous in the long run. Being daring, however, also carries risks. Increased novelty-seeking can become addictive, and while some choices can end up as mere wastes of money or time, some could be dangerous.

So I encourage new gardeners to start with the familiar old stuff, then work up to the challenging new plants or at least try new varieties of familiar plants.

Just hedge your bets. The old orange daylily and the cemetery white iris are grown all over the world—why not start with those?

Don't let fashion put you off from planting whatever catches your eye. Pace yourself, learn to think twice; it's sometimes best to just say no to every new plant!

Top Plant

This chapter simply has to end with what, in my opinion based on decades of looking in all corners, is the World's Favorite Passalong Plant. Turned out to be pretty easy.

For sure, winnowing the list of favorite plants shared among gardeners in any region is hard enough; by continent it's even dicier, but the whole world? Pushes the concept to the edge.

But of all the plants I have seen shared widely across five continents and in every kind of garden, from the finest manors and botanical gardens to humble cottage gardens and even in cemeteries, one stands out head and shoulders: *Hemerocallis fulva*—the common tawny daylily.

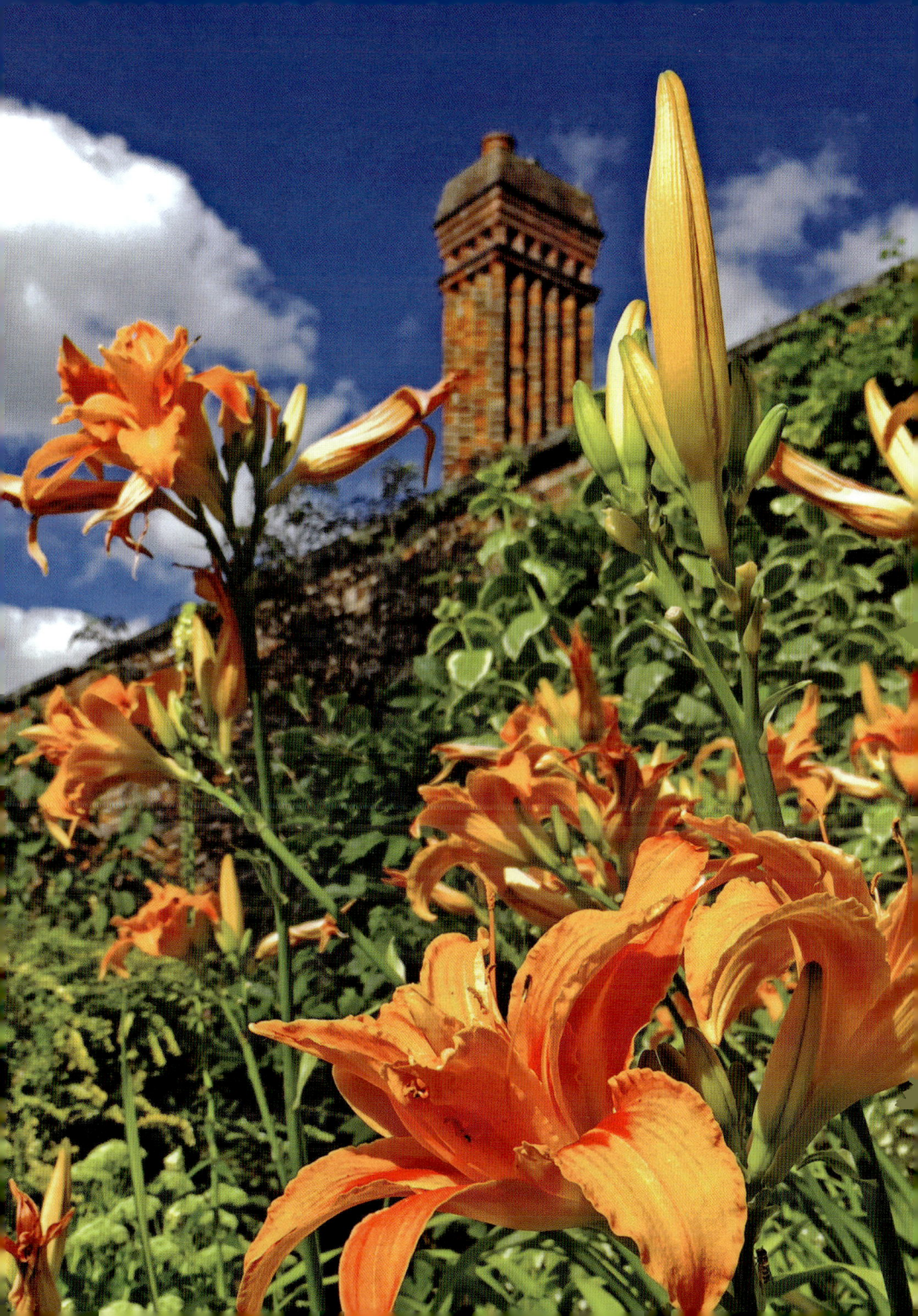

Widely derided as the outhouse lily or ditch lily, the most popular forms are the double-flowered 'Kwanso' and triple flowered 'Flore Pleno' which can have up to eighteen petal-like tepals per flower!

On top of being pretty enough to pass muster even at the Hampton Court Royal Botanic Garden, all parts are edible—the flowers in particular have the same vitamins as broccoli, (and are a lot easier to grow).

And talk about being shareable! The clump-forming edible perennial is a mule—the sterile triploid doesn't set seed, meaning every single one you see is a divided clone of the original plant grown now for over three thousand years. It depends utterly on humans to get around the world.

Like or not, when you see that ancient ditch lily, it's durable and loveable, easy to share and easy to digest. Those are good reasons to make it the World's Favorite.

To those lip-curlers who call it a weed, I say "If you can't beat it, eat it."

Yard Art: The Good, the Bad, and the Unbelievable

Gaudy is when you do something people don't like but they think you know what you are doing, and cut you some slack; tacky is when you just don't know any better, bless your heart!
—Overheard in Philadelphia Mississippi

You know there was a general discussion around the communal fire soon after the first Stone Age man or woman moved from scratching animal shapes into clammy cave walls to daubing paint on rocks outside and hanging woolly mammoth tusks from trees to make sounds in the wind. Right?

It was a prehistoric epiphany that resonates to this day, including in the garden where problems may arise because some neighbors get it, some don't.

All my life, as I navigated a world where most people obsess with being as unique as everyone else on the block, I have pondered my own urge to "gaudy stuff up"—a huge esoteric part of DIGr garden culture.

Try to understand that my 1960s Hawai'ian-shirt formative years were influenced by unusual events. Every Sunday evening Tinker Bell swirled her wand to change *The Wonderful World of Disney* opening scene from black and white to color. On Saturday afternoons a young man wearing shoes spray-painted silver with yellow lightning bolts would regularly do wheelies on his farm tractor, right down my small town's main street. Oh, and there were John Lennon's "cellophane flowers of yellow and green, towering over your head."

I was also keenly aware of the irony in how my horticulturist great-grandmother Pearl held her nose over the cheap concrete yard chicken cherished by her less educated daughter-in-law (Granny), while she herself had made a ring of homemade concrete toadstools for under her cedar tree.

As Scottish landscape artist Ian Hamilton Finlay put it, "Better than truth to materials is truth to intelligence."

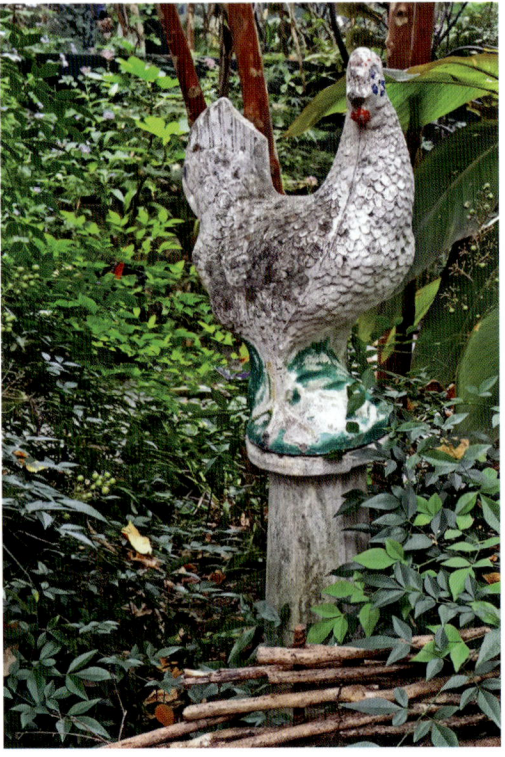

Granny's chicken in a place of honor.

I have always been a confirmed dumpster diver who drags home *objets trouvés* to pair with other cast-off treasures into funky assemblages, from bathroom fixtures and old musical instruments to weathered gnomes and homemade whirligigs (I don't care what you call it, anything with a propeller is a whirligig).

And my garden sports a dozen glass bottle trees, including one that is nearly twenty feet tall. 'Nuff said.

I've come across thousands of similar garden expressions—what I call the good, the bad, and the unbelievable—in both very fine and downright campy gardens, all over the world. Between Alaskans spraying snowed-under shrubs with green colored water, to topiary artist Pearl Fryar's outlandish South Carolina shrub sculptures and the world's oldest stumpery at Biddulph Grange in north central England, the only thing they have in common is an individual who has a vision, some time, and a lot of unused stuff to work with.

Natural Inclination Gone Wild

Bejeweled yards are little more than simple outdoor extensions of the normal urge to arrange glass figurines in a kitchen window, or to array little quaint *tchotchkes* on the patio table.

"We can do a lot of things with plants," according to my notes from a landscape class taught by Professor Neil Odenwald, "but if you're not careful, herbaceous material can all run together. All great gardens have some sort of artwork or embellishment to personalize them, to draw attention to a specific area, lead the eye from point to point, and provide a visual bridge to carry the garden through all seasons."

Usual accessories include fountains, statuary, sculpture, urns, obelisks, pottery, sundials, animal figures, wall or other hanging objects, and folk art such as scarecrows, birdhouses, stacked rocks, pink flamingos, and glass bottle trees. And those are the normal ones; I've seen countless one-offs like the woman who spray-painted her favorite NASCAR driver numbers on her elephant ears, and a Stonehenge replica made of clothes pegs.

English garden designer John Brookes wrote, "If there were ever a place to laugh, it is in a garden. To suddenly come across an amusing piece placed among vegetation or by the side of a pool is always a great bonus."

Even useful "hard features" such as gates, seating, fences, walk paving material, or night lighting fixtures can be modified to invoke a unique style that says something about your interests.

"And," Odenwald continued, "it has *nothing* to do with the quality of the piece; aged concrete can look 500 years old. And in a small space you can get a million dollars' worth of embellishment from a single well-placed clay pot."

Personal Expressions

It's a natural urge for gardeners to use symbols as interpretations of their relationship to the natural world, or to express strong feelings or ideas, or reminders of people. Yard decorations are nearly always very personal forms of expression, of taking a risk by putting ourselves on show.

Cultural historian Stephen Young explained, "It's usually not done for the purpose of being different; some of these gardeners just have a different take, and are probably very conventional in all other ways.

"We all have this urge, by the way," he continued, "only it occupies more brain space with some folks, and they express it more directly. The rest might be afraid to lighten up, lest something else begin to slip."

Often folk artists with unabashed yard shows, after being asked too many times to explain themselves, are at a loss for fancy phrases; they stammer out simple but heartfelt hints along the lines of "It's just something in me that has to come out."

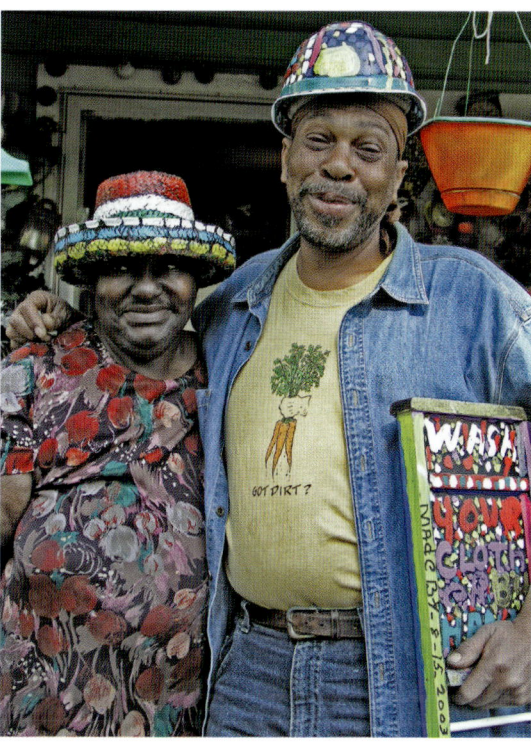

Dirt meets outsider artist L. V. Hull.

As we worked together on a village display, Earl Simmons, a folk artist from Bovina, Mississippi, struggled to explain his work, saying that what he creates "is what most people already see, only different, not like anyone else would do. It's the same thing, just put another way."

To me, the most honest was L. V. Hull, the "shoe garden lady" of Kosciusko, Mississippi. Her small fenced flower garden was bedecked with countless dozens of gaily painted shoes on sticks stuck between her flowers. Most visitors just saw the shoes, but if you took away the ornaments, underneath it all there was an everblooming, lowmaintenance flower garden, full of color and texture yearround.

Hoping to get some clues or insight, I asked L. V. what compelled her to put things out there, why does she do it? "Because," she smiled, with a gentle but emphatic wave towards her neighbors, *"I'm not like them."*

Too Much of a Good Thing

Just as coffee tables, mantles, and kitchen window sills can quickly get cluttered with knickknacks, objectophile-at-heart DIGrs tend to accumulate lots of mismatched stuff.

Some of it is designed specifically for effect. Case in point is the ubiquitous pink plastic flamingo, which has been America's "most loved to be hated" garden accessory since first patented in 1957 by art school graduate Don Featherstone. "Flamingo people" feel a connection to one another, using the ersatz birds as a sort of flag to symbolize membership of a slightly off-kilter club with no bylaws or rules—a community of people drawn together solely by their interest in doing things differently.

In one liberating flash, in a conversation we had after a shared lecture on yard art, Featherstone stated his perspective about the disdain some people have for his ubiquitous creation with a simple declaration: "Before plastic, only the wealthy could afford poor taste."

But all this is a slippery slope. I am proud to have been at London's Chelsea Flower Show in 2013 to photograph the first garden gnomes allowed at the prestigious show in its hundred-year history. And within five years there were old fashioned handcrafted bottle trees at the show. Just sayin'.

So, what goes around, comes back around. Maybe it's a sign whimsical DIGr art will eventually be accepted for what it is: Accessorizing with a message. It's in our hearts, our homes—why not in our gardens as well?

Worst thing that can happen, is people will talk about us. But hey—they've been doing it for thousands of years.

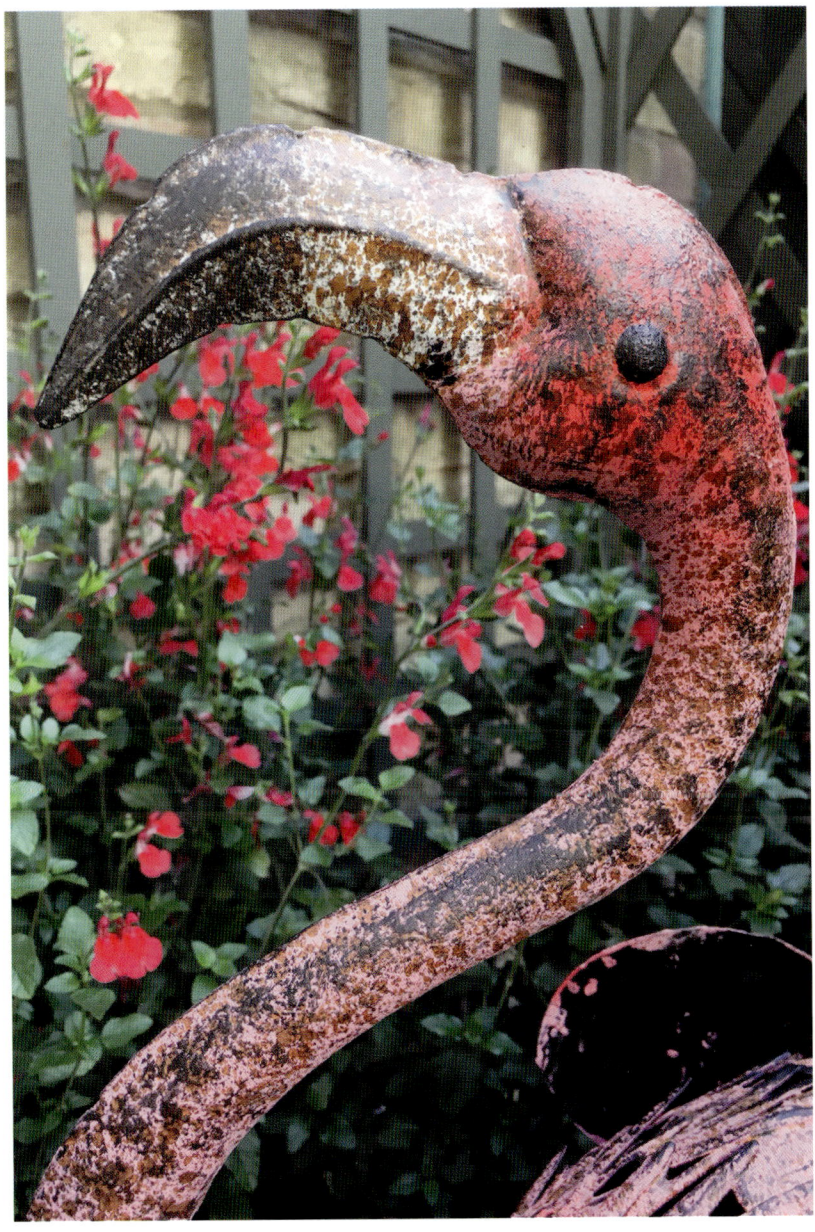

Guardians of the Garden

Garden gnomes have been around for many centuries, always with even the cheeriest looking ones raising suspicions.

These whimsical reminders of the frivolous side of gardening are small vernacular versions of the oversized naked goddesses you see in expensive, more formal gardens in which high-end designers realize the evocative nature of sculpture. Plus, like small saint statues (Fiacre, anyone?), they are more keeping in scale and mood with small gardens than would be larger statuary.

Gnomes, clad in conical top-flopping red or green Phrygian caps (which signify freedom and the pursuit of liberty), were first named as such by sixteenth-century Swiss alchemist Paracelsus from *genomos* meaning "earth dweller." Wood, porcelain, and terracotta *gartenzwerge* started circulating in the early 1800s Germany and hit Big Time when in 1847 Sir Charles Isham imported nearly two dozen to his English garden; Lampy, the lone survivor—arguably England's first immigrant gnome—is insured for a million pounds sterling (well over a million dollars).

Make a Tire Planter

Eudora Welty struck a chord in the 1940s, when she described women and children arguing over tires off a wrecked car. The children needed swings, but the women wanted them for planters.

"Crown" tires, so called because, when cut and inverted into zaftig pots, they resemble frilly crowns, are now everywhere, from Alaska to the fashionable outer reaches of Long Island and England, Japan, and all over Europe. They are, after all, as intellectually valid for containers as recycled whiskey barrels, cooking kettles, and the like.

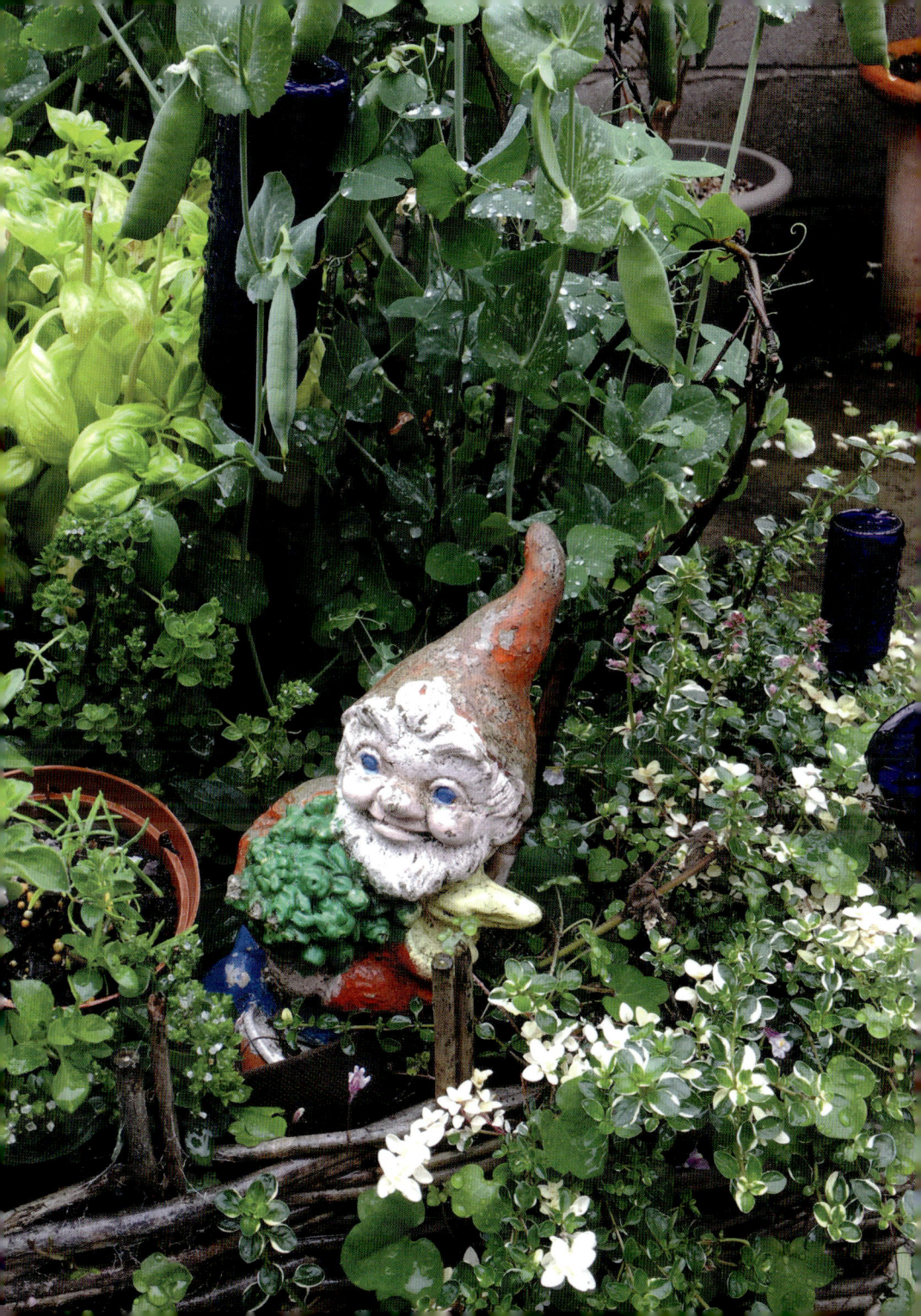

While gardeners worldwide use tires "as is" or maybe painted to make them seem more acceptable, there are three simple but important tricks to cut and invert one without herniating yourself: Choose the right kind of tire, cut just the sidewall, and practice the special move that helps turn it inside out. Skip any of these, and you will fail.

NOTE: It *does not matter* if the tire is steel belted or not. The steel wires are only in the tread, which is not cut in this process.

1. Choose your tire carefully
Lay a tire on its side, and use your foot or hand to push the curved "shoulder" in a little way. If you can't do this easily, move on to another tire.

2. Cut the tire
Any sharp knife will easily cut through the soft, thin sidewall of a tire. You might draw a pattern before cutting. NOTE: Cut away from yourself in case the knife slips (trust me on this!).

3. Invert the Tire
You can't out-brute a tire, you gotta out-think it.

Stand the tire up, cut side away, and with your knee push in the "shoulder" (remember step one) while pulling back on the cut edge with your hands. Shift your knee a bit until you find the right spot.

4. Work your way around
Pull and push the rest of the tire, *a little at a time*, all the way around.

Paint and then hang the leftover "sun" on a fence or wall.

You can use the cut-out part as a wall decoration, or as a mower or string trimmer guard around shrubs.

By the way, whether or not a tire has a rim, *there is no difference* in the steps taken; if the tire has a rim on it before you start, it will end up looking like an urn with pedestal.

If you want to paint your tire planter, hit it first with a household degreaser, take it to a car wash and use the soapy spray and then rinse (don't wax). ANY kind of paint will work, but fast-drying spray paint is easiest.

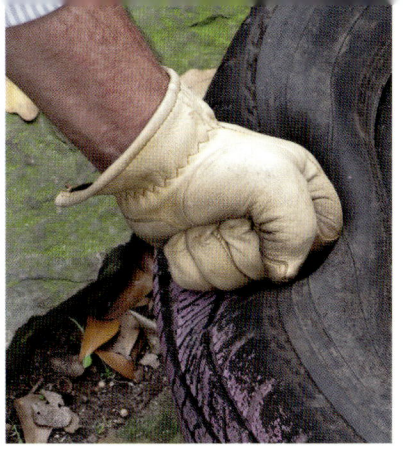
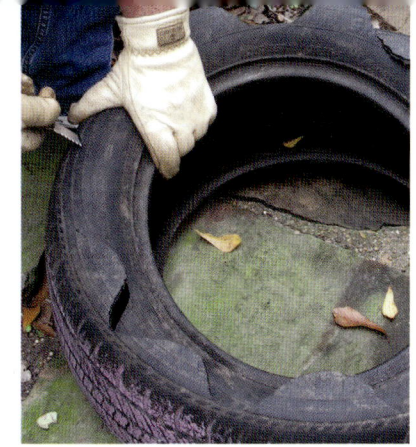
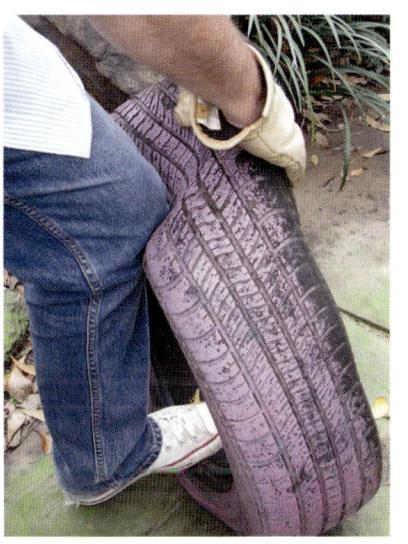
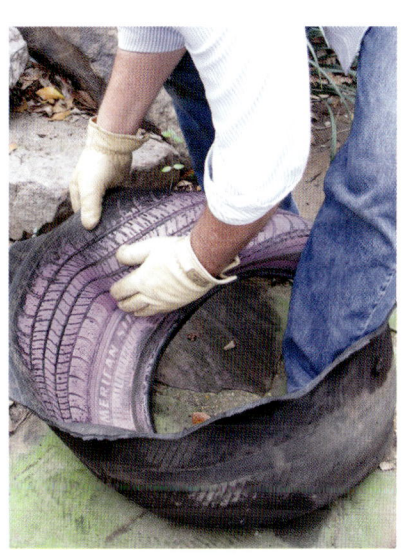
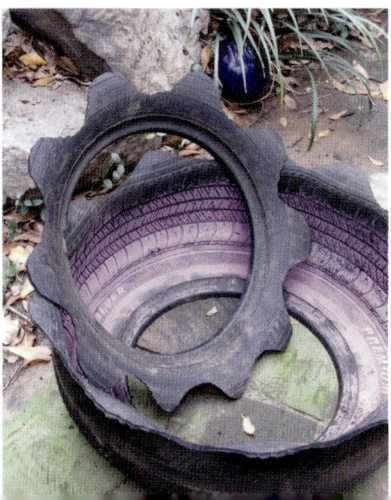
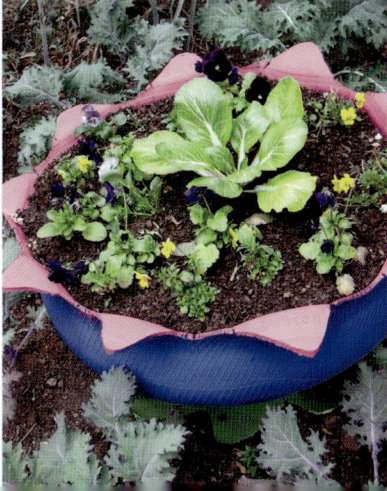

Bottle Trees—Poor Man's Stained Glass

Some folks say a glass bottle tree is a forgettable sight, but from the time I saw my first one I knew my own garden could never be complete without at least one. In my mind, a colorful bottle tree is hands-down the most awesome personalized garden eye-poke possible.

Not all garden glass is tacky. High-end glass sculptures, from Dale Chihuly's fantastic creations to a trio of large locally made plates of art glass suspended in my garden, are popular features in many great gardens. But simple homemade glass bottle trees—what I call

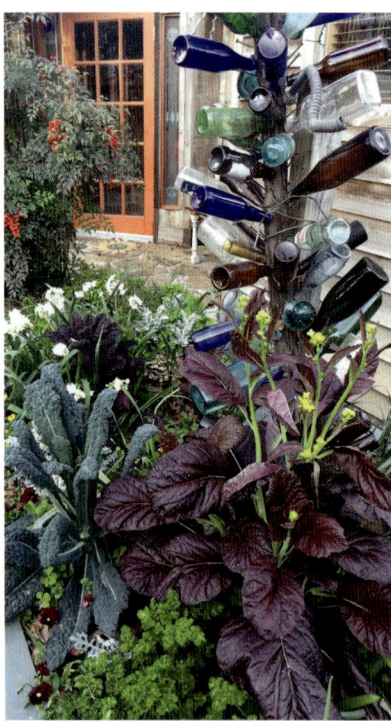 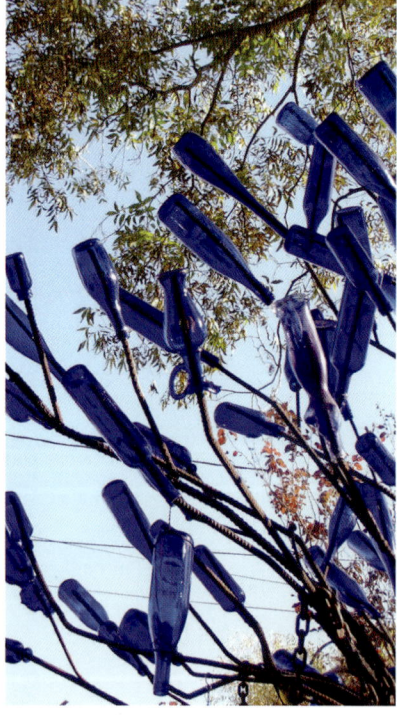

DIGrs' stained glass.

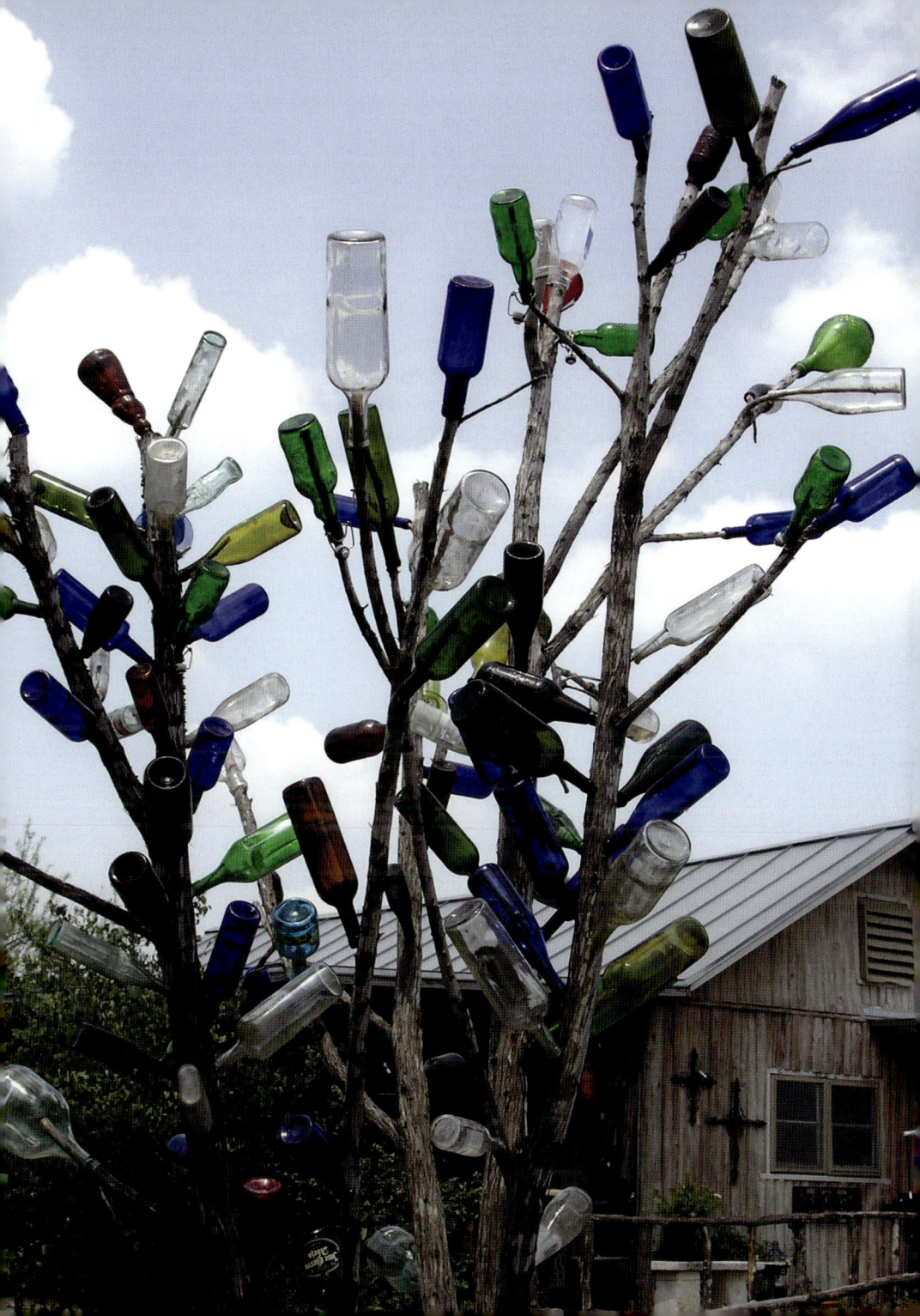

"redneck Chihulys"—often appear in major flower shows, including Royal Horticulture Society extravaganzas in England.

Bottle tree construction is a *concept*, not a recipe, every single one a variation of the simple theme: bottles stuck onto or hung from something in the garden.

And no two are alike—each *Silica transparencii* is a unique combination of colors and shapes glinting in the sun, which represents a gardener displaying shameless confidence in who he is.

They can be made of wood with nails, small dead trees, big limbs tied to posts, welded metal rods, rebar rods stuck in the ground, or upended pitch forks, then festooned with colorful bottles or other hangable glass.

What's this about evil spirits? From when the first hollow glass bottles began appearing two thousand years ago in Egypt and Mesopotamia, Arabian tales (remember Aladdin?) began to circulate about the genies and spirits that resided within—probably originating by bottles, like crude Helmholtz resonators, making moaning sounds when wind blew over their openings.

Somewhere in there, people started thinking bottles could handle bad spirits. The idea was, roaming night spirits would be lured into and trapped in bottles placed around entryways, and morning light would destroy them.

Whatever. But one genuine quality of bottle trees is how, like a row of bottles perched in the kitchen window, their cheerful colors may stimulate winter-withered pineal glands, helping stave off Seasonal Affective Disorder (winter blues), which certainly lifts our own spirits.

Besides, it isn't about haints, it's about garden color. As English glass and metal artist Jenny Pickford told me, "All we're doing is holding glass up to the sky so its colors can sing."

It isn't scary to put up a bottle tree. It's scarier not to be able to put up a bottle tree.

The Godwottery

Little did he know at the time, but Thomas Edward Brown, a late nineteenth-century Victorian poet from the Isle of Man, put a word to the kind of garden many DIGrs have: *godwottery*.

While composing a very short poem simply titled "My Garden," he threw in a little-used word that means "knows" as a rhyme for "rose plot" and "fern grot," a shortened word for grotto:

"A garden is a lovesome thing, God wot
Rose plot, fringed pool, ferned grot..."

Now godwottery has come to mean an overly elaborate garden with a wild assortment of plants and objects such as shrubs pruned into whimsical spirals, succulents planted in a toy truck, windmills, flowery cast-metal furniture, fake rocks, gnomes... lots of gnomes...

You know. A bodacious DIGr garden.

And it's a good thing, god wot!

Dementia Concretia

Finally, there can be a problem here. Some folks pass the tipping point with garden art, to the point where they can't seem to stop.

Those blithe souls are sometimes affected with a mild version of more serious impulse control disorders like rubbish hoarding (*syllogomania*); their *dementia concretia* syndrome is an excessive compulsion to build unconventional structures, sculptures, or figures using whatever materials are readily available—usually concrete, bottles, cans, scrap metal, and other industrial and household junk—and they keep going way beyond where others would stop.

Outstanding examples, including the late Elmer Long's bottle tree forest in the Mojave Desert, the Ava Maria grotto in Alabama, and

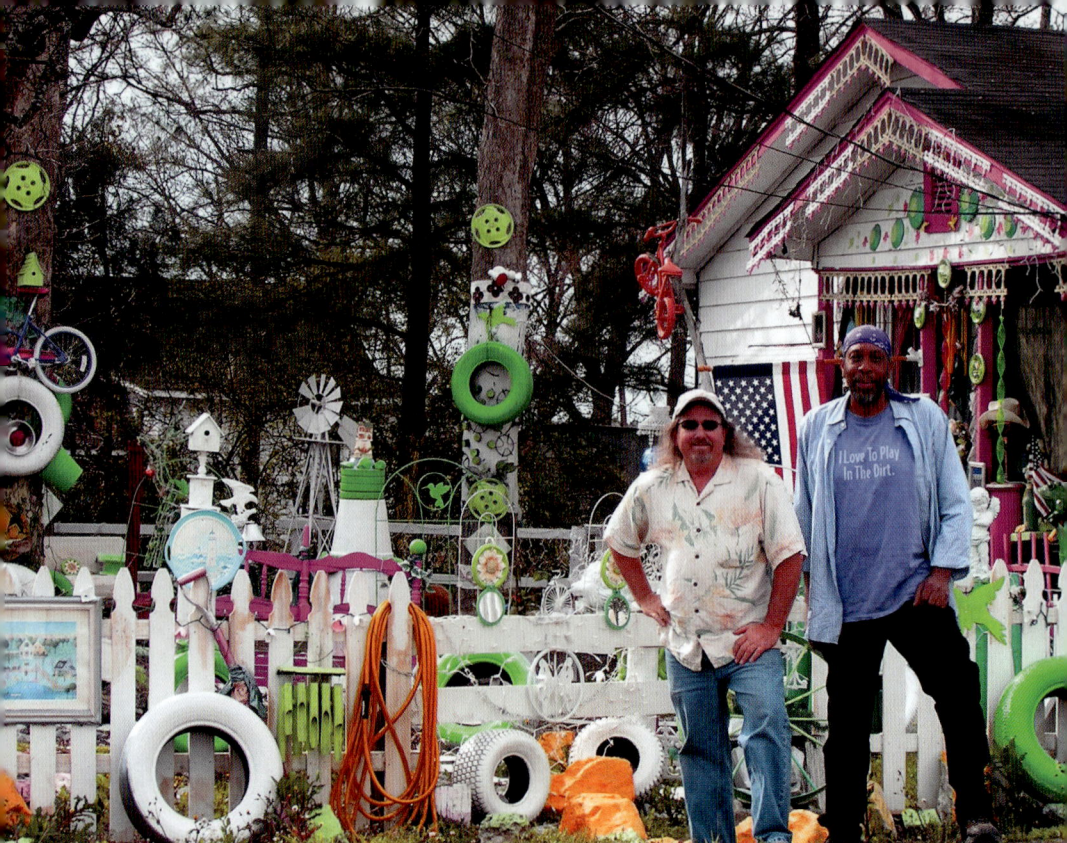

Mississippi Delta yard show.

Spanish architect Gaudi's over-the-top constructions, are sometimes preserved as folk or outsider art. However, most of the time, when the original creators pass on, so do the creations.

My brother once told me that there's a fine line between sane and crazy, but I think it's a wide gray band. Just remember what Earl Simmons said, which is for you to "do what pleases yourself because most folks ain't gonna agree with you—you just have to hold up your head and go on."

In the end, art is for everyone. But because one person's perfume is another's stink, we won't always see eye to eye. My idea of good, bad, and unbelievable will be different than yours, or anyone else reading this—if you pass the book along, the recipient will have a different idea too.

Perhaps DIGrs just have a wider band of tolerance for what they consider unbelievable.

7

Putting It Together

Gardens are not made by singing "Oh how beautiful" and sitting in the shade . . .
—Voltaire

I'm not getting into much practical instruction in this book because in spite of huge advances in techniques and materials, the simple act of gardening never has been rocket science.

It's been around since the first Neolithic human observed an above-average bean and saw the value in showing someone else how to save and replant seeds.

Besides, I bet you already have a bookshelf groaning under books chockablock with reinvented traditional how-to routines and variations. A lot of garden-variety gardening is just "making do" like Rumpelstiltskin weaving golden garments from common straw.

Whether it's a pollinator-buzzing wildflower meadow, floriferous cottage border, windowsill brimming with cascading plants, or a kitchen garden of colorful vegetables and fragrant herbs, fanciful creations are built from the dirt up, using whatever can be gleaned locally.

And, like eating an elephant, it can only be done one bite at a time.

With this in mind, here are a few DIGr concepts and tricks of the trade worth considering, in no particular order. Grab a pencil or highlight pen, and take your time.

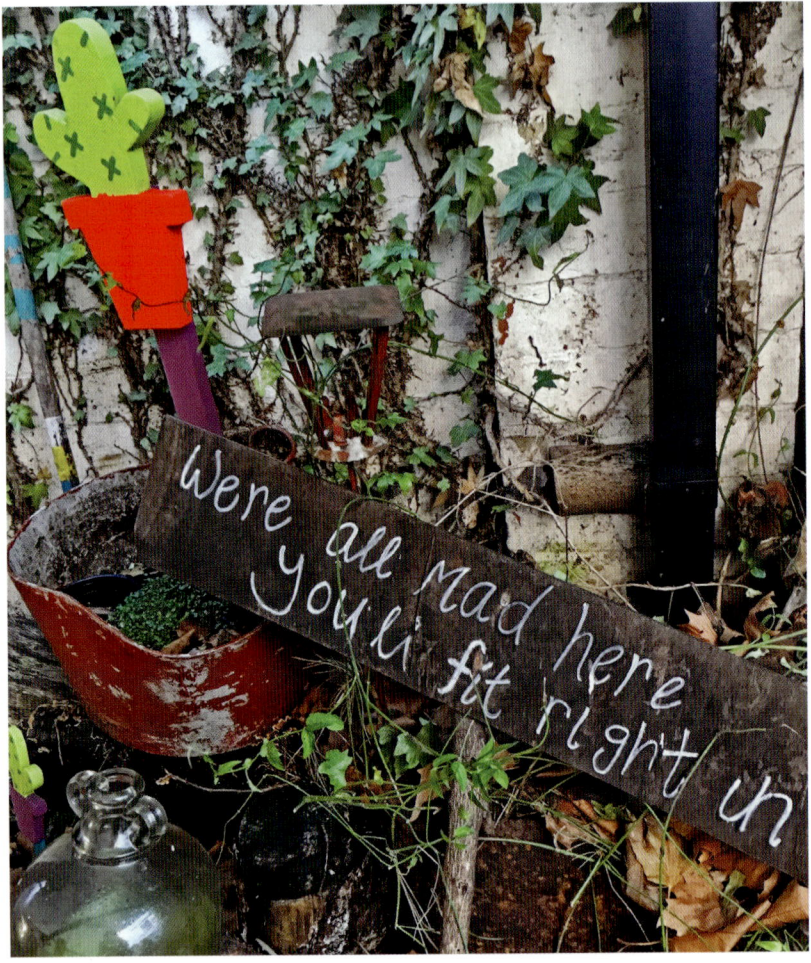

Gardening 101: Ignore Your Inhibitions

Pardon the deconstructionism, but all gardeners basically do it the same way: Dig a hole or fill a pot, put a plant in it green side up, give it a modicum of nurturing, protect it from troubles, and take it from there. If it fails, try again or move on to another plant.

Everything else is a variation on that theme.

If you are clueless or nervous, the easiest approach when getting started is to try learning the way kids do. The old "can't teach an old dog a new trick" has been debunked, and there are benefits from taking on childlike attitudes. Kids expect to make mistakes, but instead of saying, "I can't do that," they think, "I haven't learned that yet."

The keys to learning quickly are simple: a desire to garden, making a start, and rolling with mistakes.

Start with knowing why you need to grow stuff. Do it because you want to, not to impress anyone. Keep it simple and relevant—Yard of the Month may not be as doable or rewarding as, say, growing herbs you can add to a simple meal or planting zinnias to attract butterflies or prettying up the house with posies.

Early on you will run into difficulties (happens to the best of us) but it's important to understand that it's okay to not know it all. This is the way real gardeners feel and think, and even laugh about with other gardeners. The best disasters often become legends (world-famous rose breeder David Austin's first-ever tray of rose seedlings died).

Finally, watch other gardeners. Let them know you are learning, and don't want to get overwhelmed, and just want to ask a question or two. You'll be surprised how readily others will not just stop what they are doing, but actually keep going and involve you, too. Note: Learn when to say, "I've had enough for now."

Let go of your grownup inhibitions, and jump in playfully—like a child. The more often you do this, the bigger your comfort zone gets and the more at ease you will become.

Feel like you deserve accolades? Make yourself an award sign. Or tell folks what you really think.

Smart Gardening

The American Horticulture Society has long promoted a checklist of guiding principles for good gardening, all of which are part and parcel of being a DIGr. Collectively called the SmartGarden™ tenets, a few of them are

Know Yourself. Figure out how much time you have for gardening, what you want to grow, your willingness to try new ideas, and your physical and financial limitations.
Adapt When Necessary. Improve soil with organic matter, cut trees where you need sunshine, plant shade where you want shade plants. Make the path less muddy.
Pick the Best Plants for Your Conditions. Avoid impulse buying; choose mostly those that can grow well in your garden and with your personal commitment to gardening (if you don't like mowing, don't have a big lawn).
Have Fun! It's okay to have a little whimsy in your personal garden, to express yourself.

Recologists

DIGrs are notoriously adept at the practice of recology—the science of 'again' which originated in the ecology mantra of "Reduce, Reuse, Repurpose, and Recycle."

It's as simple as running leaves, spent flowers, kitchen scraps, and even weeds through a leaf pile or compost bin, then returning the nutrients to the soil. Plant in anything cast away that can hold soil. Cut old plastic jugs to cover and protect tender seedlings or rose cuttings from frosts, or poke tiny holes in the bottoms and sink into the ground beside vegetables to water plants slowly. Line flower beds with wine bottles, car tags, and broken dishes.

Whatever. We're just not gonna just throw perfectly useful stuff away.

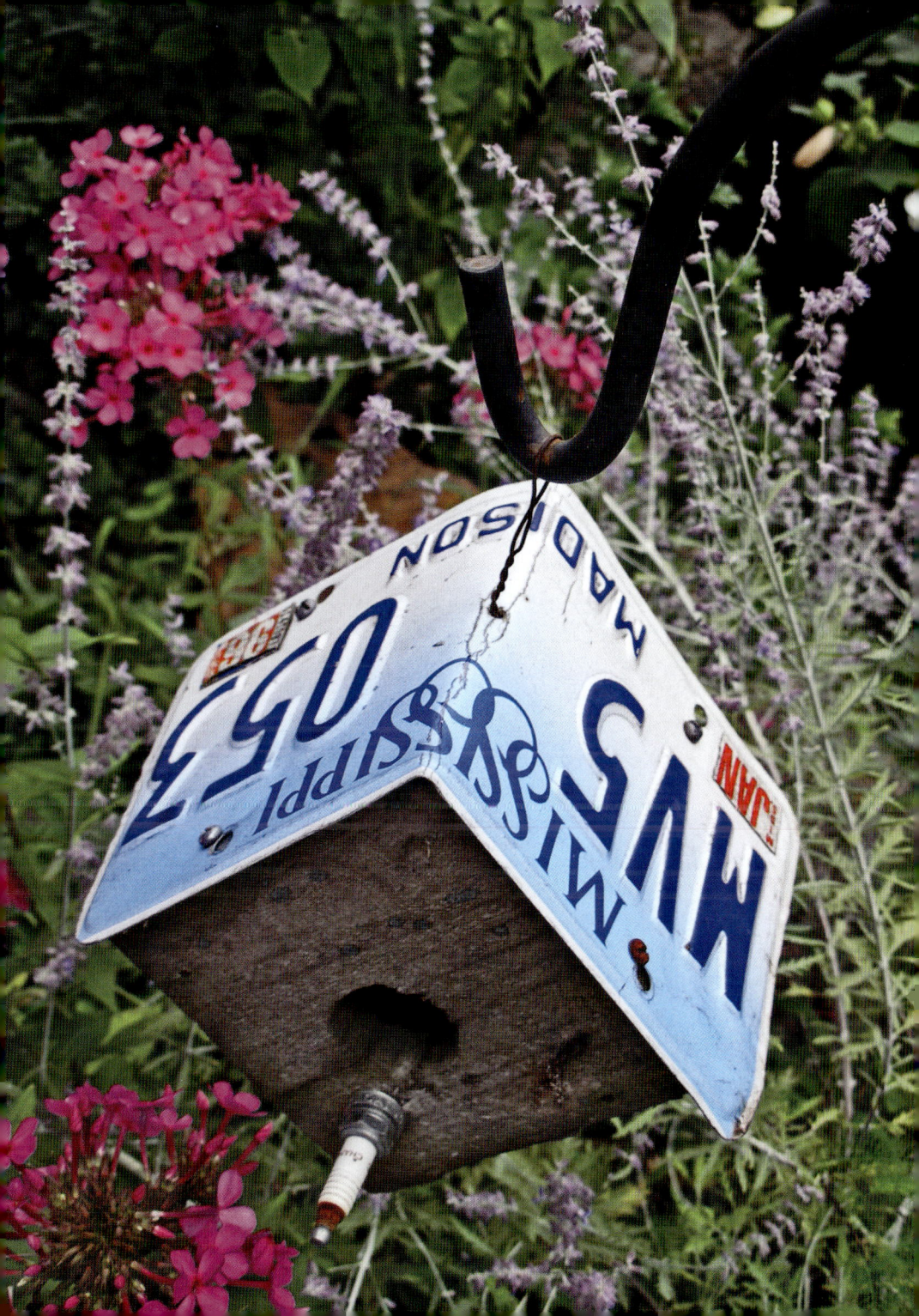

Compost Knows

Healthy plants need healthy soil, and the best way to improve your native dirt is by adding stuff to it, usually some sort of organic material such as bark, compost, leaves, or the like.

Think of it as being like adding a few crumbled crackers to a bowl of chili, and you have both the idea and a general understanding of the proportions.

So why not just grow your own compost? Don't get bogged down in how-to, just remember—*compost knows what to do*. Simply pile up leaves, grass clippings, vegetable scraps, spent flowers, etc., and give the heap time to decompose into useable stuff all by itself. Happens in the woods all the time.

You can speed nature up with all sorts of tricks like chopping the debris into small bits, keeping it moist, adding extra nitrogen, and turning it over and over. I've made finished compost in as little as three weeks, but after working my butt off I realized that *nobody cared*. It ain't a race, it's a process.

Now I follow three simple rules: Stop throwing all that stuff away, pile it up somewhere, and give it time. When I need compost, I dig into my leaf pile, sift the bottom stuff through a sheet of half inch hardware cloth into a wheelbarrow, and throw the unfinished stuff back onto the pile.

Don't overthink or feel confined by well-meant but basically dumb rules, like "don't put weeds or meat into the compost." Hmmph. I put all that in my leaf pile, even a dead raccoon one time (buried pretty deep, of course); it all composted just fine.

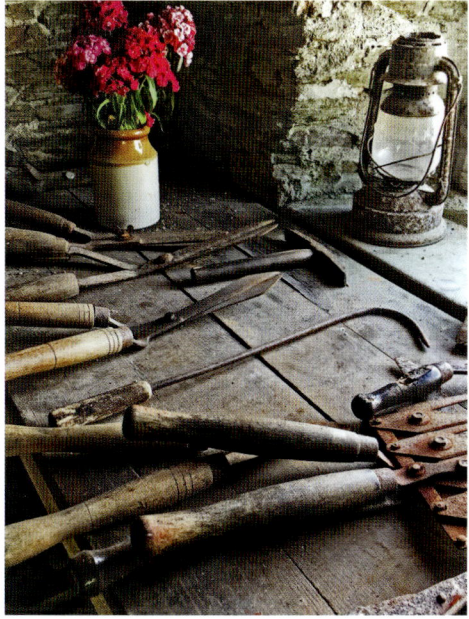

Quintessential Tools

Some tools are so simply useful and easy to use that nothing can take their place. If we didn't have 'em, we'd invent them.

While peeping into old garden sheds I often covet tools worn smooth with use. Gardens with heritage, such as Hidcote, proudly display them like an heirloom art gallery.

DIGrs usually leave their tools where they can be reached easily (and tripped over by unwary visitors); I'm no exception. My most-used tools include

- Garden fork I can use to work the soil without killing worms
- Flat-bladed spade for digging; scoop shovel for moving dirt
- Flat file for keeping blades sharp so they work better

- Trowel for planting annuals without tearing up my old hands
- Pruning shears and a scabbard to protect my pocket
- Curved pruning saw for limbs and small shrubs and trees
- Five-gallon buckets for hauling, mixing, and storing stuff
- Hatchet for splitting firewood and hammering stuff
- Iron poker for rearranging burning logs in my fire pit
- Rain gauge just to satisfy my curiosity

If you are a DIGr, you got some others. Right?

Critters

Birds and butterflies and bees—*oh, my*! You won't find a DIGr garden without critters, both good and the not-so-welcome.

Nowadays everyone agrees that pollinators are important; however, they aren't all here just to grace our gardens with beautiful entertainment and salve our eco-consciences. Some, especially bees, big spiders, diverse reptiles, raccoons, opossums, rabbits, deer, moles, voles, squirrels, and other wildlife creep in from time to time. Add in all the insects, snails, and other plant-eating creepy-crawlies that some of the larger wildlife feed on, and the chain of wildlife can quickly get problematic.

But they are part of the web, and the best approach is to ignore them as best you can, using mechanical controls (fence for deer, traps for smaller animals, netting or squashing for bugs) when practical and resorting to poisons only in specific instances where losing a plant or a pet is likely.

I mean, you can't have lady beetles without aphids for them to eat. And don't be too smug about your butterflies, without tolerating a few plant-eating caterpillars (which are essentially just hungry teenage butterflies).

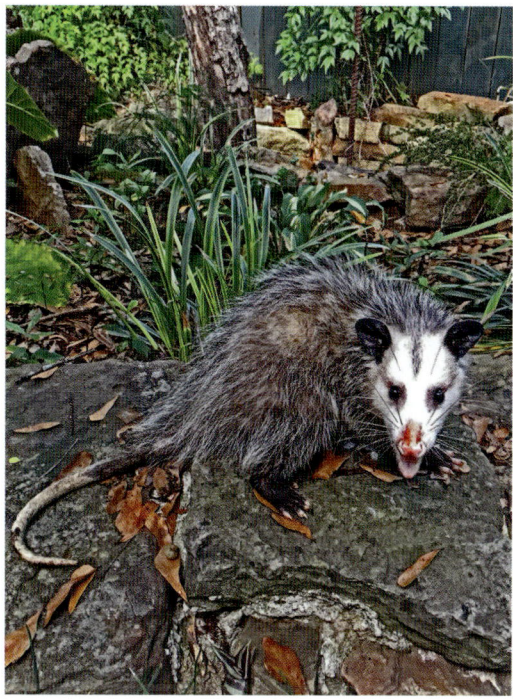

It's a bug-eat-bug world out there, and good gardeners learn to live and let live. Within reason of course.

Wildflowers in Suburbia

I have a neighbor—a DIGr wannabe, if you ask me—who wants a slice of wildflower meadow, bees and butterflies included, in her front yard. And still get along with the neighbors.

No mean feat, without jumping through a few hoops.

The typical "suburban foundation" style of planting was originally designed to make quickly built houses packed into sterile new suburbs

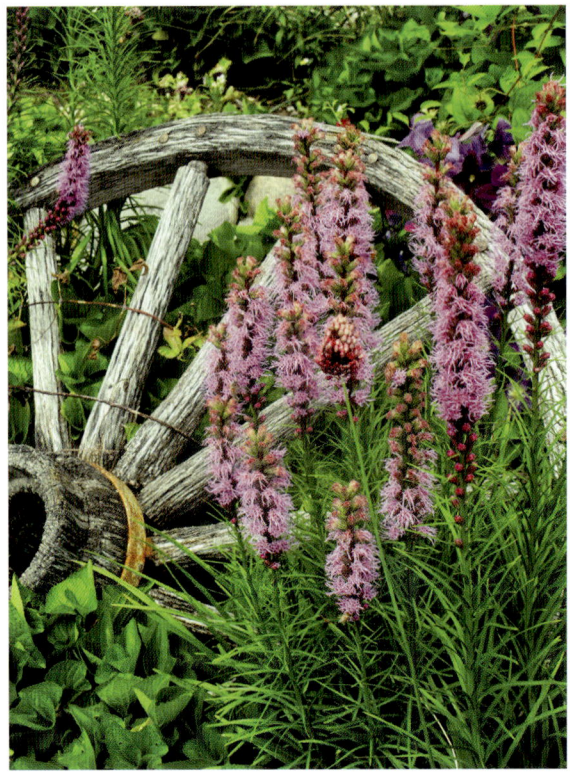

A rustic hard feature helps interpret wildflowers.

look less like cracker boxes. It's like tucking parsley around a roasted pig, but ya gotta do what ya gotta do to appease the neighbors.

Five things she can do to get away with this unruly approach:

- Keep some lawn, even just a postage stamp area, and have some of it go alongside the front curb. Keep it mowed.
- Carefully site a rustic hard feature to provide interest; a group of posts topped with birdhouses, a short section of split rail fence, a large piece of

- driftwood or small boulder, a bird bath—whatever looks kinda naturalistic but is still nice.
- Plant a few "normal" flowers—daylilies and the like—and add a few well-accepted native perennials such as purple coneflower, summer Rudbeckia, tall purple verbena, and gaura. You know, stuff you can get at garden centers all the time that are publicly accepted as "regular" flowers.
- Smile and wave at the neighbors. Involve them slowly by showing them a butterfly or getting them to look really close at an individual flower to see the incredible details such as the "rabbit face" in a larkspur flower.
- Important big step: Go online to the National Wildlife Federation site, fill out the form, and have them send you a metal Backyard Wildlife sign. It will alert neighbors that you belong to a larger movement and are getting expert advice guiding everything.

Then, once these things are pulled off, gradually slip in the wilder plants a few at a time, all the while smiling and waving at the neighbors—they'll never suspect a thing.

DIGr Garden Routines

Doing things as needed, rather than because we're supposed to, works.

Earnest gardeners often recommend unnecessary things, and regularly spread erroneous information, usually based on flawed research, cherry-picked details taken out of context, or downright hocus-pocus lore. Or they read it online so it must be true.

I understand that commercial horticulturists have different motivations than home gardeners, and that for every "This always/never works" there is always someone out there who swears otherwise. But DIGrs know that, in general, with a nod to occasional exceptions:

- You don't *really* have to have your soil tested, just don't overfertilize. And compost doesn't need lime.
- Roses don't have to be pruned above five-leaflet leaves.
- Planting by the moon is . . . well, it doesn't hurt to do it.
- Marigolds don't actually repel bugs (they attract spider mites).
- There are practically no truly dependable deer-proof plants.
- It's okay to pollard crape myrtles and other small trees.
- Pots don't need rocks in the bottom for better drainage.
- Nonnatives are often as good or better for pollinators.
- Native plants can be among the weediest in the garden.
- Picked tomatoes don't ripen, they just get soft and redder.
- The color wheel isn't set in concrete.
- The occasional, careful use of chemicals won't kill you.
- Vinegar doesn't kill weeds, it just burns them back awhile.
- You can reuse potting soil, and don't have to sterilize pots.
- Compost piles don't have to be turned regularly.
- Double digging is a masochistic futile effort in rainy areas.
- There is no pie in the sky or rock candy mountain . . . Sorry.

No Winter Blues

Autumn is exhilarating with its flurry of colorful falling leaves and the idea of no grass to mow or shrubs to prune until spring. To most folks, winter gardens are mostly structural garden bones of trees and shrubs with different shapes in shades of brown and green, and it's okay.

But DIGrs have spent years finding plants that have interesting foliage, berries, bark, and even flowers in the depth of winter.

This is of course dependent on where they live, but if you just look around there's always something—even where it snows a lot. Winter birds are at their most colorful and entertaining, keeping everyone alert that life is still happening outside the fogged-up window.

Plus, DIGr gardens are all but overrun with "hard" features—arbors, fences, gates, water garden, and even compost bins—that not only help the landscape function, but also give structure. Sometimes a patio table is the only thing that indicates how deep the snow is!

Deep winter is when garden art shines, piquing attention when "normalarian" gardeners are moaning about their struggling holiday poinsettia losing its colorful bracts indoors. DIGrs act ahead of time so when the first really cold night falls, their gardens will still work, freeing them up for other things like bringing in potted plants and starting a bowl of paperwhite bulbs to bloom on the kitchen counter.

Umami and the Smell of Fresh Dirt

One thing DIGrs have a corner on is savoring little things. Focusing on the "here and now" enriches garden experiences that often comes in surprising forms.

Just going outside to put birdseed in a feeder, or brushing through a rosemary shrub leaning over the walk, are whole-sense exercises. Soon as you step outside the temperature, humidity, and the fresh air will hit your skin, and you'll see shapes, colors, and motions. Your ears will pick out the splashy water garden, the whine of mosquitoes, and the chirrup of an evening gecko. There's the smell of cut grass or fresh mulch or *something*, and you may even be able to nibble on a few herbs, edible flowers, or fresh raw vegetables.

All sorts of amazing stuff come into play; Jenny Nelson walks around in her bare feet, which helps her experience ever-changing textures, temperatures, and more.

A classic overlooked example is, though we dig in it all the time, we don't stop to think about what gives dirt its peculiar odor.

You may already know that good soils smell kinda nice (if you do know, you're a DIGr). Not talking about the low-oxygen stink of wet

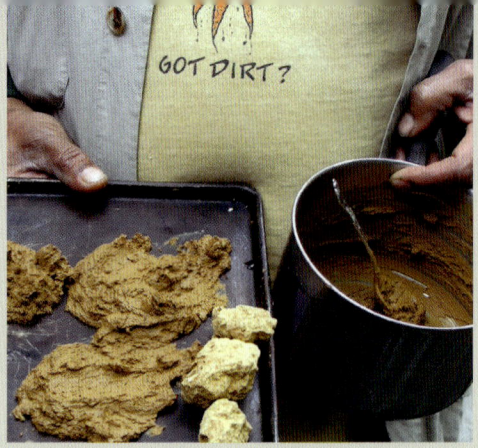

Mud Pies

I spend an almost obsessive time digging in my little garden, turning it over and over as a form of therapy. But other than getting a little grit when chewing a torn fingernail, I normally wouldn't eat it.

On one of our weekly radio programs Dirt and I got a medical researcher to explain the benefits behind geophagy, the ancient but still-practiced concept of people eating clay for missing minerals, which can buffer against acidic foods or reduce the effects of diarrhea.

That last one is obvious; Kaopectate, an over-the-counter diarrhea medicine, is a slurry of chalky kaolin clay from Georgia. But before you go gnoshing on dirt, check with a doctor—eating too much can cause digestive disorders and keep the body from absorbing iron, which can lead to anemia.

Dirt eaters prefer chalky-textured clays. But because they are usually bland and taste, well, earthy (kinda gritty, too), adding vinegar gives it a tangy ring.

Dirt talked about how he prepared his and told where the right kinds could be found. Here's his recipe:

- Find some light-colored clay (the color of coffee with cream, usually along a hillside).
- Let it dry, then break it up into a cup.
- Add 3 tablespoons vinegar and a pinch of salt.
- Add enough water to make it like thick mud.
- Spread on a cookie sheet and bake in a slow oven until hard (this also sterilizes it).
- Break into pieces, store in plastic bags. Munch as you feel the need.

blue mud, or the sweet pong of fresh manure; it's the slightly sweet fragrance you get when you turn compost or smell a rain coming.

Vaguely similar to catfish, lake water, mushrooms, and freshly dug potatoes, this smell comes from an oil called geosmin that's exuded by bacteria; when raindrops hit dry soil, or as a low pressure front ahead of a storm moves in and "degasses" the soil, small bubbles of geosmin float to the surface and release aerosols called petrichor—which is what we smell.

There may not be a great word that captures these extra sensory experiences, so I just go with *umami*, which is usually associated with food as the savory flavor that complements those of sweet, salty, sour, and bitter. Think mushrooms, poached fish, chicken soup, and other hard-to-pinpoint tastes, and you have umami.

Garden umami, then, is how I sum up the ability to gather the full meal deal of sensory gardening, beyond the usual five physical senses.

Grow Your Own

Sharing can't happen without . . . well, giving away bits of your plants by seed, division of bulbs or crowns, rooted cuttings, and a few oddball others like those plants that grow on the ends of leaves.

While there are now many good online sources for heirloom seeds and bulbs, in general, DIGrs save and share their own, waiting for seedpods of vegetables and flowers like zinnia, celosia, kale, and okra to fully mature so the seeds are dry, then run them through a kitchen colander to remove fluff and insects.

Their gardens usually have a place set aside for rows of pots with divided plants, or jars of cuttings rooting in water.

Main thing is, plants want to be shared, but they depend on us to get them from garden to garden. Practice on easy stuff, and take it from there. Go to a plant swap for even more practical tips. Better yet, get to know a nearby DIGr.

Plant Swap: Organized Chaos

Some years ago over supper, celebrated folk writer Eudora Welty told me that her mother "stopped going to her garden club meetings when they stopped swapping plants."

But not to worry, the tradition is still alive with plant swaps from tiny Flora, Mississippi, to Sheffield, England.

On prescribed days, dozens of men, women, and children gather, hauling pots, buckets, egg cartons, trays, and plastic bags of homegrown shrubs, vines, seedling trees, bulbs, seeds, tropical plants, wildflowers, vegetables, and herbs. All pass the Passalong Triarchy test, and are ready to go to new homes and gardens.

The sheer variety of plants brought in over the decades has been astounding. Though some are quite rare, almost all were at one time commonly grown but are now nearly impossible to find for sale anywhere. Most have little labels and homemade signs stuck in their pots with sweet folk names and general care advice.

Once the plants are all lined up, participants run their eyes over the assorted treasures and decide on a plant to hope for. To hope hard for.

The mechanics of large swaps are simple. The plants are set in rows or in groups according to use (vegetables and herbs, indoor plants, shrubs, perennials, etc.), and at a given time people either mingle and grab, or take numbers and pick the plants that have corresponding numbers on their pots.

The organizers of PlantSwap UK, the oldest continuous plant swap in England, start their informal swap with the announcement to "get what you want, not too much, be courteous, and have fun." And everyone pretty much does just that, with lots of convivial swapping and talking, and just enjoying being at ease with other gardeners. It's a relaxed, informal mix of unusual plants and unusual people, and it works.

The one at the Flora library, which for decades has hosted the longest-running swap in the known universe, is a bit different. Each plant is given a number when brought in and lined up. At the beginning of

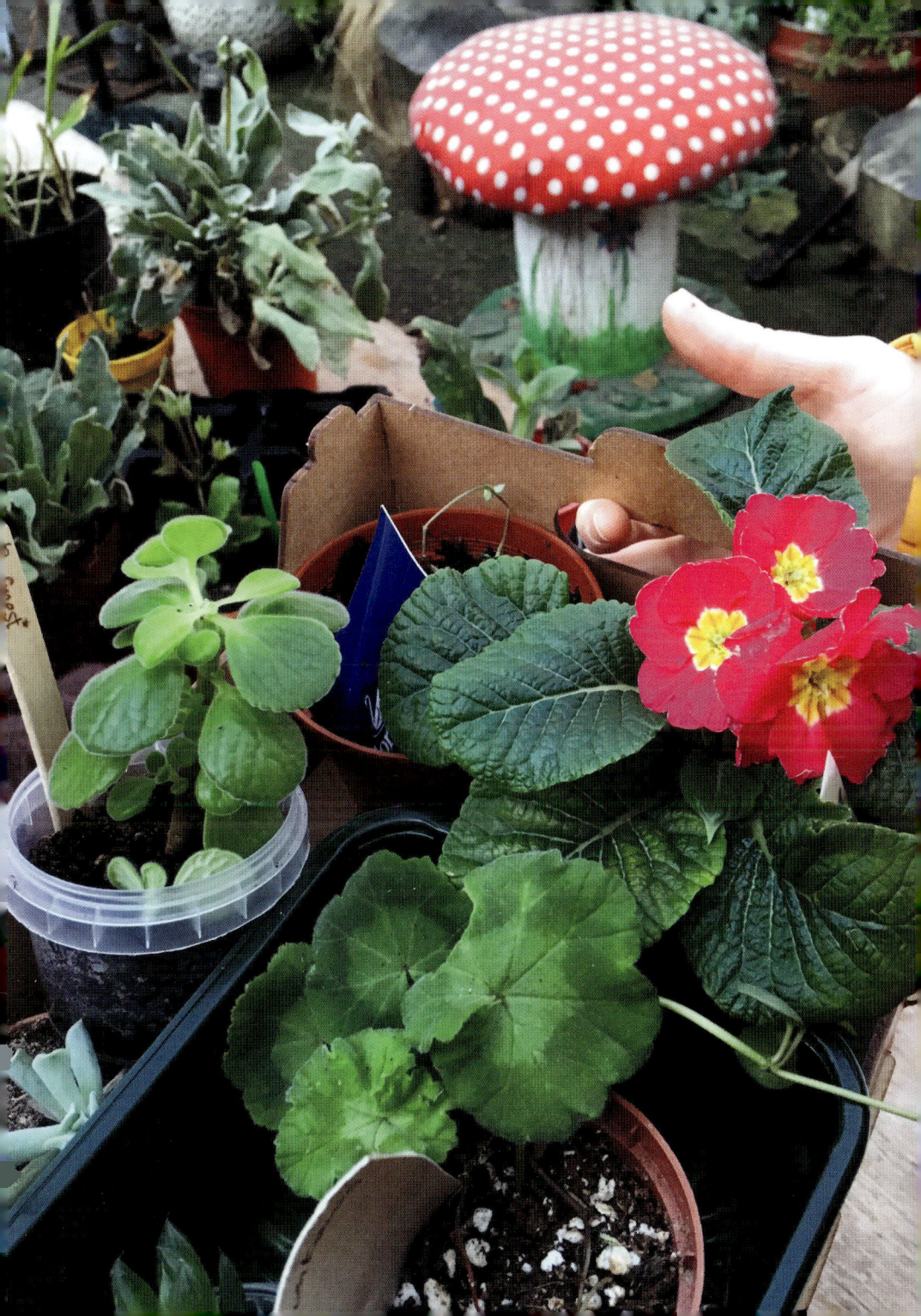

the swap someone talks a bit about the plant diversity, maybe asks about particular plants (who brought it, what do you call it, how does it grow, will it freeze outside, etc.). Then a basket is passed around from which participants draw numbers to see which plant they get—like it or not, grow it or not. The real swapping goes on later in the parking lot.

These horticultural free-for-alls simply put diverse plants and people together and mix 'em up in a safe, friendly setting. It's a DIGrs day out.

DIGr Ponderings

Finally, it comes to this: DIGrs don't just piddle around our gardens for the day; like everyone else, we often wonder what Life is all about, and deal with the doubt of if or how our efforts may impact the world.

Yet we fervently believe that, in an endless then-and-back-again loop, our ever-reproducing plants and their lore are propelled through the generations, making our efforts mean something down the line.

In a letter to her elderly brother in February 1963, my horticulturist great-grandmother Pearl wistfully described her garden nibbida; she was seventy-five years old at the time, and was eating fig preserves that she got me, as a ten-year-old boy, to help put up the summer before:

"I guess it is a sure mark of old age but I am surprised at myself for the way I have changed in the last few years. The things I used to value I care nothing for now . . .

"Flowers, especially jonquils, used to almost intoxicate me. I never got too sick to mull in delight over flower catalogues and dreaming over what I might do with some new variety. In traveling, my head was always hanging out the car window scanning the landscape for new flower acquaintances and perhaps to collect some new treasure.

"And sometimes I came up with valuable new things of my own—a white Rudbeckia, a shrimp-pink wild iris, 328 seedling chrysanthemums. Where are they today? Mostly gone with the seasons.

"Since gardening is recommended as a fine occupation for the aging, it is possible that other things contributed to my loss of interest. I'll name a few. One, being the big amount of work required. Two, the lack of interest shown by my family, giving emphasis to the thought and feeling, 'What's the use? Nobody to care for them when I am gone.' Three, the dead end I reached in the garden club . . . to buck the selfish self-seekers cost more than they were worth.

"But maybe the real answer is that I had simply become surfeited. When springtime comes and my beloved jonquils start blooming again, I may scrape up a little interest, though it may not be the consuming fire of old."

Here it is, over half a century later, and her jonquils are spread across the countryside, still blooming, as are her chrysanthemums, figs, and fig preserves recipe. And I am fortunate to find myself passing them and their lore along to others, including to Pearl's great-great-grandchildren.

From there, who knows?

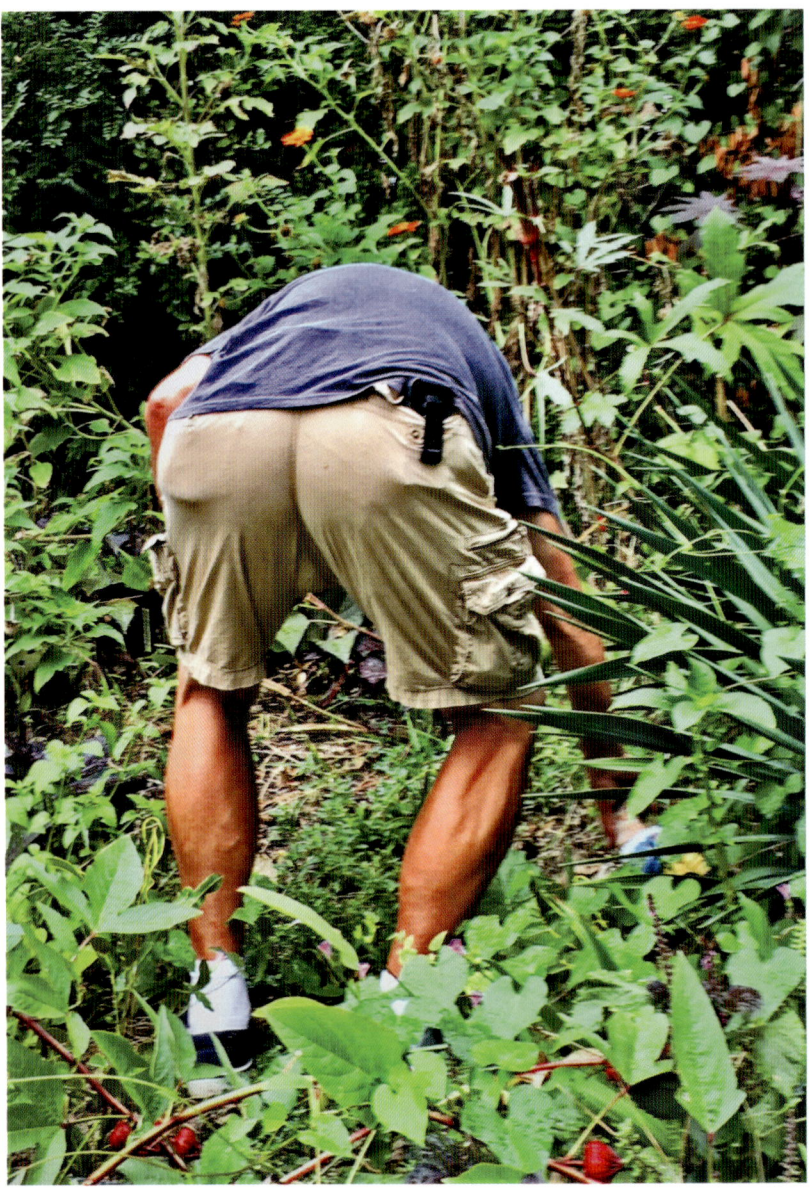

Coda: DIGrs Infinity

And be you fruitful, and bring forth abundantly in the earth, and multiply therein . . .
—From the book of Genesis

DIGrs, in their unexpected ways, offer others a confidence boost to brandish a little personality. They're showing that gardening is not a team sport with refereed rules; it is a big tent under which everyone has opportunities to shine.

Their garden lifestyle is simply about living intense versions of themselves, and hoping for understanding.

Just as you wouldn't dream of going into someone else's house to turn their toilet roll round the other way, let's honor our different approaches.

Resolve from today to follow your bliss, get excited, look closely, and experiment. Share what you grow and know with others, especially young people.

And hold your head up high—it'll work out fine.

Empowering Terms for DIGrs

> These little things may not seem like much but after a while they take you off on a direction where you may be a long way off from what other people have been thinking about.
> —Sir Roger Penrose

Here are some thought-provoking terms and ideas used in the book, with brief definitions. Most are disarmingly commonplace; all offer insights into where DIGrs are coming from.

Do you need to know them? Nah. Most of you already feel them (see Assimilation). But glance through them every now and then, and tap into what helps DIGrs cope.

Absurdism—finding humor or calling out ridiculous behavior when faced with unreasonable concepts or actions or things; examples: calling a plant either good or bad, when it's just a plant; hearing someone say a saint statue is okay but a gnome is not

Amateur—someone who loves gardening for pleasure, not profit; one for whom drawing smiley faces on too-late-to-ripen tomatoes makes it all worthwhile

Antique—an object of a certain age; with garden accessories it is generally at least fifty years; see Heirloom

Assimilation—change in behavior or thoughts after a new concept or activity is finally understood

Bohemian—socially unconventional or nonconforming person usually involved in the arts (including gardening)

Busy-body—a meddling, prying, or gossipy person; the *bêtes noires* of DIGrs

Chore—anything that *has* to be done, like it or not; see Sisyphus

Cognitive Dissonance—feelings and thoughts when an "either this or that" situation requires a hard choice between equal opposing actions

Cultural bias—the phenomenon of interpreting and judging activities or others by standards inherent to one's own culture; example: not cutting someone slack for being different

Dementia concretia—difficulty or inability to cease compulsive making, building, or planting

Dogma—a principle or set of often suspect principles laid down by an authority as incontrovertibly true

Egalitarian—pursuit of life, liberty, and happiness based on an equal consideration of interests (can't we all just get along?)

Epiphany—a sudden and striking realization; a "Eureka!" moment

Esoteric—insider views designed or understood by select few, examples: swapping potting soil recipes, tricks to growing large pumpkin, racing to get the earliest tomato

Existentialism—individuals being independent and responsible with their own values rather than whatever roles, stereotypes, or other preconceived categories they may fit ("I'm okay, you're okay.")

Fast food gardening—the use of mass-produced plants and materials for instant gratification, with little personal interaction

Feng shui (*fung-schway*)—ancient intuitive concept of positioning garden elements for a peaceful, harmonious balance

Freaking out—the sudden, startling casting off of restricting standards

Garden—a planned, people-centric space of any size, indoors or out, where plants are cultivated and displayed alongside manmade materials

Gaudy—deliberately unrestrained flair; overly colorful or frilly; can apply even to high-end objects used in less-worthy settings

Gemütlichkeit (*guh-MUTE-li-chite*)—the cozy, genial atmosphere of a place where people feel welcome and accepted

Gestalt—"pattern" or "whole" rather than parts taken out of context

Gobsmacked—utterly astonished, overwhelmed with surprise

Godwottery—whimsically overplanted and over-accessorized garden

Green thumb—natural ability to notice, make predictions, and act on subtle differences in plant growth, weather, other garden events; citing a lack of this is the most common bad excuse to not engage in gardening

Groupthink—enforced harmony through conforming to established norms without critically evaluating the alternatives

Guerilla gardening—cultivating ornamental or edible plants on neglected or abandoned land not legally available for such use

Heirloom—an inherited or otherwise shared plant, tool, or object; can be any age; see Antique

Hügelkultur (*HOOG-l-culture*)—raised beds or planting mounds made with piled logs and limbs topped with leaves and compost or dirt

Hygge (*hue-guh*)—the feeling of coziness, intimacy, peace, relaxation, togetherness; the peaceful pursuit of everyday happiness

Inscape—the unique, distinctive design or nature of every single thing (every garden has unique qualities making it incomparable to others)

Joie de vivre (*jwah de VEEvre*)—exuberant enjoyment of life; found by following your bliss

Knowledge of Disenchantment—knowing that objects of desire will likely be unsatisfactory; "this plant probably won't grow here"

Koten engei—attaching high value to plants with deliberate or bizarre mutations; opposite of uniform beauty (ugly dog contest)

Laissez faire—attitude of letting things take their own course without interference ("live and let live")

Left brain—analytical, logical, or sequential thinking or actions; examines parts of the whole in a rational objective manner

Lingo—local or "insider" style of verbal communication, such as calling soil "dirt" and using folk names for plants; DIGrs are okay with this

Locavore—person who seeks out locally grown or produced plants or products

Make-do—improvising, using tools or materials at hand

Nibbida—complete disenchantment, aversion, and weariness with things we are taught to appreciate

Objets trouvés (*ObJET trew-VAY*)—"found" objects, often used in creating vernacular yard art

Outlier—a person whose style or approach is highly atypical and stands out dramatically from that of the general population

Passalong Triarchy—the three requirements for a plant to be shared between gardeners: possess values, easy to grow, easy to propagate

Pedantic—pompous attention to academic knowledge and formal rules

Per aspera ad astra—through difficulties to the stars!

Permaculture—using ancient practical systems such as composting, rainwater collection, and natural pesticides for sustainable gardening

Phenology—the study of regular plant and animal life cycles and how these are influenced by seasonal variations and relate to garden activities

Plant blindness—overfamiliarity with a situation leading to inability to notice common details (can't see the trees for the forest)

Plant purgatory—a queue of assorted pots of plants waiting, often in vain, to be planted

Plant zones—artificial system of categorizing plants according to narrow environmental conditions; not accurate due to many variables

Play—imaginative activity done for pleasure or whimsy, without regard to reality or consequences; the way DIGrs learn

Potager garden—a small mixed garden of vegetables, herbs, and flowers, usually associated with the kitchen

Quintessence—quality of being perfectly apt or useful, often very simple and usually can't be improved upon

Recology—the science of reusing, repurposing, recycling

Refrainer's remorse—regret following having not acquired or done (or planted) something desirable

Right brain—intuitive and subjective, creative, and feeling; considers the whole of something, including intent, over the individual parts

Roony-poot—self- or elder-taught; no formal horticultural training but understands old ways of gardening; associated with "green thumb"

Rumpelstiltskin—fairytale character who wove golden garments using common straw (similar to what DIGrs do with ordinary plants)

Schadenfreude—secret pleasure derived from the mistakes or misfortunes of others; the snicker after saying "I TOLD you not to do that!"

Self-transcendent—the peaceful state of being an integral "as one" part of the garden, beyond material needs; accepting what is here and now; usually expresses itself through humility, divergent thinking; Nirvana

Sisyphus—mythological human doomed to repeatedly and eternally roll a rock up a mountain; ceaseless, unnecessary labor (i.e., lawn mowing)

Slow gardening—thoughtful, sensual pursuit of gardening bliss through all seasons, using local, sustainable practices and sharing with others

Speciesism—assigning different values or rights on the basis of species (weed vs. wildflower, good bug/bad bug)

***Sui generis* status**—of its own kind, in a class by itself (DIGrs)

Susurrus—a nearly imperceptible "almost" sound (snowfall, leaves rustling on the ground, wind sighing in trees; "white noise")

Tacky—lack of finesse in expression; the enjoyment of cheap or overused objects. Usually associated with "just don't know better"

Tchotchkes (*CHOT-skees*)—small evocative bric-a-brac items, collectable or otherwise

Teachable moment—a sudden, informal opportunity to show someone something in the garden that may make a huge difference in their life

Terroir (*ter-ROR*)—the unique character of plants created by highly local soil types, climates, and other environmental factors

Thinking disposition—human tendency to be open-minded and curious, to seek connections and explanations; set goals, make and execute plans, and question the status quo

Umami—savory; the fifth taste beyond sweet, sour, salty, and bitter; may be applied to a feeling of satisfaction

Vernacular—informal, locally common

Wabi-sabi—appreciation of asymmetry, incompleteness, and impermanence; celebrating the passage of time and intransience of life

Weed—any plant having to deal with an unhappy human

About the Author

Felder Rushing is an eleventh-generation American gardener, a non-stuffy horticulturist who travels the world looking for simple garden approaches, which he promotes in his newspaper columns, books, magazine articles, and NPR radio program. The author of over twenty books and founder of Slow Gardening, he was named by *Southern Living* as one of "25 people most likely to change the South."